新工科英语系列教程

Academic English for Petroleum

石油学科英语

主编　王新博　付晓　徐小雁

清华大学出版社
北京

内容简介

本教材是基于新工科对教育资源供给“同频共振”的要求，以专门用途英语（English for specific purposes）教学理论为指导而编写的石油学科英语教程，旨在满足能源类高校大学生及行业从业人员提升国际学术交流能力的需求。本教材以石油人文为主线，分八章阐释石油学科的人文内涵及其对国计民生的重大影响，具体包括：“石油与民生”“石油与经济”“石油与政治”“石油与科技”“石油与环保”“石油与教育”“石油与文化”和“石油与新能源”，为能源类高校大学生及行业从业人员进入相关工程领域学习和工作奠定基础。

本教材适用于高校和行业领域学术英语教学的需要，满足高校国际化人才培养对学术英语交流能力的需求；对于具备高中英语水平或相当基础的广大英语爱好者而言，也具有重要的参考使用价值。

图书在版编目（CIP）数据

石油学科英语 / 王新博，付晓，徐小雁主编. — 北京：清华大学出版社，2019
（新工科英语系列教程）
ISBN 978-7-302-53774-8

Ⅰ. ①石… Ⅱ. ①王… ②付… ③徐… Ⅲ. ①石油工程—英语—高等学校—教材 Ⅳ. ①TE

中国版本图书馆 CIP 数据核字（2019）第 199874 号

责任编辑： 刘细珍
装帧设计： 陈国熙
责任校对： 王凤芝
责任印制： 宋　林

出版发行： 清华大学出版社
网　址： http://www.tup.com.cn，http://www.wqbook.com
地　址： 北京清华大学学研大厦 A 座　**邮　编：** 100084
社 总 机： 010-62770175　**邮　购：** 010-62786544
投稿与读者服务： 010-62776969，c-service@tup.tsinghua.edu.cn
质量反馈： 010-62772015，zhiliang@tup.tsinghua.edu.cn
印 装 者： 北京密云胶印厂
经　　销： 全国新华书店
开　　本： 185mm × 260mm　**印　张：** 11.25　**字　数：** 266 千字
版　　次： 2019 年 10 月第 1 版　**印　次：** 2019 年 10 月第 1 次印刷
定　　价： 49.00 元

产品编号： 085226-01

Preface 前言

《石油学科英语》是在教育部提出"新工科"建设的大背景下，为"加快培养适应和引领新一轮科技革命和产业变革的卓越工程科技人才"而编写的专用学术英语教材，供能源类高校大学生提升国际学术交流能力使用，旨在培养学生基于学科知识进行学术英语交流的能力。具体而言，该教材不仅培养学生专业学习所需的听说读写能力，还有针对性地培养其掌握科学的研究方法，有效开展信息检索并进行分类、整合、评价和推理的能力，基于数据支撑进行书面和口头表达的能力以及良好的批判性思维能力。

最新版的《大学英语教学指南》强调"将特定的学科内容与语言教学目标相结合"，"期待能提供与学科相结合的一些'桥梁性'课程"，[1] 顺利帮助学生进入专业领域学习，达到用英语进行学术交流的目的。学科英语是高校公共外语教学的必然归宿，用英语交流学科知识是体现公共外语教学的最大价值所在。"由于不同学科的语言呈现出不同的语篇体裁结构，从而反映出不同的语言特征，具体表现在语法结构、词汇选择、语气情态、文体风格等方面。"[2] 因此，专业学术交流使用的英语非普通英语，而是和学科融合的专门用途英语。高校要推进"双一流"建设，必须做好国际化育人工作，做好国际合作与交流，开展高水平国际科研合作，培养一大批能在学科领域参与国际学术交流的高素质人才。国际化人才培养的需求最终要以学科英语为支撑才能实现，这也吻合《大学英语教学指南》的要求："要将特定的学科内容与语言教学目标相结合，教学活动着重解决学生学科知识学习过程中所遇到的语言问题，以培养与专业相关的英语能力为教学重点。"可以说《石油学科英语》的编写恰逢其时，不仅顺应新工科建设对教学资源供给"升级换代"的时代要求，还满足用英语进行国际学术交流，汲取本学科国际前沿信息的迫切需要。

---

1. 贾国栋. 2017.《大学英语教学指南》与高校大学英语教学改革. 当代外语研究，62－65.
2. 曾蕾，尚康康. 2018. 学术英语教学与学科英语研究的互动模式探讨. 西安外国语大学学报，53－59.

本教材以石油人文为主线，围绕石油学科这一主题，阐释八个方面的内容，即“石油与民生”“石油与经济”“石油与政治”“石油与科技”“石油与环保”“石油与教育”“石油与文化”和“石油与新能源”，多角度展示能源领域的合作、交流和发展。各章自成体系，既有各自的内嵌逻辑关联，又有章节间的衔接和照应。各章从国际权威杂志或网站精心选取了大量主题鲜明、逻辑严谨、表达清晰、结论可靠的学术文章，培养学生汲取、整合、归纳和评价信息的能力和基于文献的书面和口头的学术英语交流能力。每章都由单元学习目标（Unit Objectives）、课前活动（Pre-Class Activities）、课中活动 (While-Class Activities)、课后活动（After-Class Activities）和语言综合训练（Integrated Exercises）五个部分构成。

单元学习目标（Unit Objectives）

指学生学习每单元后要达成的目标要求，以任务的形式展开，主要包括了解石油学科某一领域的最新发展、该领域的热点或焦点问题以及围绕问题或主题进行的一些思考、评价或预测等。

课前活动（Pre-Class Activities）

主要是为课内教学活动做铺垫和“预热”准备，以背景知识导入为主，通常包括查阅核心术语文献、检索和整合学科基本知识以及补充和完善背景知识等。该部分主要要求学生以小组为单位合作开展信息检索并整合相关信息。

课中活动（While-Class Activities）

主要围绕单元核心主题并结合学生课前活动的已有认知图式，开展信息整合、分析和评价，权威专家的观点解读及自我认知的剖析和评判等活动，通常包括学生小组展示、观点辩论、论点归纳、自我认知评判、小组总结等形式。

课后活动（After-Class Activities）

主要引导学生围绕课内学习过程中的难点、痛点或争议点进一步补充和完善相关信息并开展小型学术调研活动，通常包括权威专家的观点解读、不同学者的观点碰撞和分享、难点文献的再检索、自我学习的评价总结等。该部分主要采用观看视频录像、查阅文献、写总结或心得等形式。

语言综合训练（Integrated Exercises）

考虑到学科英语的语篇修辞特征，该部分在设计上没有沿用常规语言强化训练的编写方式，而是采用以应用与产出为导向，强调在语境中获取学术信息的习题形式。第一题是核心词汇练习，帮助学生有针对性地快速识别需要掌握的高频词汇，要求学生通过自主学习获取词汇在文本语境中的意义。第二题是核心术语积累，学生能否得体地认知核心术语将直接影响其学科知识的学习效果。该题旨在让学生通过文本阅读有效获取术语信息和文本信息。第三题是文本长难句解读，是学科英语能力检测的重要方面，旨在通过训练培养学生有效解构文本的能力，提升学生面对复杂语言环境时快速、准确获取信息的能力。

本教材在编写过程中，参考了大量国外能源领域的权威网站、学术期刊、出版书籍和杂志等，在此无法一一列出，谨向原文作者表示最诚挚的谢意和最崇高的敬意。这些作者在本学科领域的真知灼见开阔了编者的视野，让我们受益匪浅。

本教材在成书过程中得到了上海市高校大学英语教学指导委员会主任、全国专门用途英语教学专业委员会会长、中国学术英语教学研究会会长、复旦大学外国语言文学学院教授、博士生导师蔡基刚教授的大力支持，他为本教材编写提供了纲领性指导。美国俄克拉荷马大学终身教授、中国石油大学（华东）杰出校友邬星儒教授为本教材编写提供了大量的文献资料并提出了很多建设性的建议。在此，向两位专家学者表示最诚挚的谢意。

我校英语专业 2016 级、2017 级以及储运与建筑工程学院、经济管理学院 2017 级 A27 和 A34 班本科生在试用本教材的过程中提出了很多宝贵意见和建议。2018 级翻译硕士专业笔译方向的研究生马腾同学协助校对部分文稿。本教材能得以付梓还要感谢清华大学出版社各位领导的大力支持和编辑的辛勤付出。在此，向以上对本教材给予支持、关心和帮助的各位领导、同事、学者和广大学子表示最衷心的感谢。

由于编者水平有限，书中疏漏和不足之处在所难免，恳请业内同行和广大读者不吝赐教。

编者

2019 年 5 月 28 日

Contents 目录

Unit 1
Petroleum and Livelihood

Unit 2
Petroleum and Economy

Contents 目录

Unit 7
Petroleum and Culture

Contents 目录

Unit 8
Petroleum and Renewable Energy

Unit 1
Petroleum and Livelihood

Unit Objectives

Goal 1

Get well informed of what petroleum can do for human needs through task-based activities.

Goal 2

Understand why petroleum is so essential to human livelihood through cooperative learning.

Goal 3

Make an in-depth study of certain heatedly-discussed questions about the impact of petroleum on livelihood.

Petroleum does more than just provide gasoline for cars and jet fuel for airliners. Products and by-products of petroleum end up in items used daily around the world—billions of pounds of polyethylene plastic alone go towards making plastic bags, food containers, hula hoops, and other consumer products.

Pre-Class Activities

Activity One Get to Know the Role of Petroleum

Search online to make a list of petroleum-derived products and then group them into different categories. Share your categorized list with your team members.

____

____

____

____

Activity Two Get to Know the Significance of Petroleum

Read the following article carefully, summarize the key points about what petroleum can do for human needs, and prepare for a class discussion.

Reading A

Petroleum: A Commodity Essential to Our Very Way of Life[1]

Edward Cross

President of Kansas Independent Oil and Gas Association

❶ When talking about petroleum, many people likely have the image of a barrel filled with a thick, black substance. But what most people may not realize is that petroleum is the building block of thousands of products that make our lives more comfortable, safer, cleaner, and healthier.

❷ When thinking about the role of oil and gas in our lives, many people may only think of

1. Cross, E. (unknown publication date). *Petroleum: A commodity essential to our very way of life.* Retrieved from https://www.kioga.org/public-information/op-ed/petroleum

vehicles and fuel, but petroleum plays an integral role in nearly every aspect of our lives. As a matter of fact, over 6,000 products come from petroleum. People use oil-based products every day, whether it is your television remote, cell phone, or even the toothpaste and toothbrush you use to brush your teeth.

❸ Synthetic fabrics such as nylon and polyester are made from petroleum. In addition, as a key component in heart valves, seat belts, helmets, life vests, and even Kevlar, petroleum is saving tens of thousands of lives daily. Furthermore, oil and gas are key components in many medicines and antibiotics, such as antiseptics, antihistamines, aspirin, and sulfa drugs.

❹ These are just a few of the improvements that oil and gas make in our lives and societies around the world, and as a top 10 oil producing state, Kansas is a major contributor to that. As a nation, we take great pride in our agriculture sector and the role it plays in feeding people around the world. In the same way, we should take great pride in the role our oil and gas industry plays in providing a commodity essential to our very way of life. Oil and gas are fundamental to our modern way of life and high standard of living.

❺ Also, the oil and gas industry makes our environment far safer and creates new resources out of raw materials.

❻ The energy we get from oil and gas is particularly valuable for protecting ourselves from the climate. The climate is always changing, whether mankind influences that change or not. In the last 80 years as CO_2 emissions have risen from an atmospheric concentration of 0.03% to 0.04%, climate related deaths have declined by 98%. Oil and gas make the planet dramatically safer.

❼ According to the Environmental Protection Agency, oil and gas methane emissions account for only 3.63% of total U.S. greenhouse gas emissions. Methane emissions from the oil and gas sector declined 3.8% in 2018, marking the fourth consecutive year of decline. The fact is our nation's 21st century oil and gas renaissance has made domestically produced oil and gas economical and abundant. This market-driven success has helped our nation achieve significant emissions reductions. The men and women of the oil and gas industry reject the stale mindset of last century's thinking peddled by some that oil and gas production and environmental stewardship are not compatible.

❽ Oil and gas have also made the planet dramatically richer in resources. Until the Industrial Revolution, there were almost no energy resources. Oil and gas are not naturally resources. Those who first discovered how to convert oil and gas into energy weren't depleting a resource; they were creating a resource. The world is a better place for it. Life is all about taking materials in nature and creatively turning them into useful resources. And by creating the best form of energy resources, the oil and gas industry helps every other industry more efficiently create every other type of resource.

❾ More than a billion people around the world face challenges for adequate food and education, clean water, and protection from heat and cold due to a lack of access to safe, affordable, and reliable energy. We should work to ensure more people have access to safe, affordable, and reliable energy, no matter which state, nation, or continent they reside in, because people need more energy, not less, to rise out of poverty and enjoy health and safety.

While-Class Activities

Activity One Enrich Your Knowledge Cooperatively

Hold a group discussion about the categorized list of petroleum-derived products you've found before class. Each group will be asked to make a class presentation to share your list.

Activity Two Review Your Findings Collaboratively

In what ways is petroleum essential to our way of life? You are required to make a brief summary by following Reading A again. You may supply additional information other than Reading A.

Activity Three Map out the Consequences

Since petroleum is so deeply involved with our livelihood, can you imagine what would happen if oil someday is running out? Read the following article, hold a group discussion about the consequences of the absence of oil supply, and make a class presentation.

Reading B

Imagining a World Without Oil[1]

Steve Hallett and John Wright

❶ Dismantle the oil rigs and stack them in a pile. Radio the tankers and order them back to port. Pull out the drills and cement up the wells. (A year after the BP spill in the Gulf of Mexico, let's hope we've learned how to do that, at least.) Tow the platforms back to shore. Plug up the

1. Hallett, S., & Wright, J. (2011, April 21). Imagining a world without oil. *The Washington Post*. Retrieved from https://www.washingtonpost.com/opinions/imagining-a-world-without-oil/2011/04/12/AFppFHKE_story.html

pipelines. And lock up the Strategic Petroleum Reserve while you're at it—it has only about a month or so worth of oil in it, anyway.

❷ What would happen next? How would we live in a world without oil?

❸ First, there's transportation. With the overwhelming majority of the oil we produce and import devoted to powering our cars, motorcycles, trucks, trains, and planes, the impact on getting around would be most dramatic. Price-gouging would begin right away, and long lines would form at gas stations. The lines wouldn't last, though, because the gasoline would soon be gone. A strategic reserve of finished petroleum products—gasoline, diesel, and aviation fuel—has often been suggested but never created. Within a month, every fuel tank would be dry, all our gauge needles would point to "E", and the roads, rails, and skies would be virtually empty.

❹ How far is it to the nearest grocery store? How long does it take to walk—or bike, or skate—to work? Finally confronting our dependence on motor vehicles, we'd reach for whatever solutions we could find. Soon, we'd all be looking for an electric car (but there are precious few of those for sale) or converting our vehicles to run on natural gas. But we'd be waiting for some time to secure adequate natural gas supplies, establish delivery infrastructure, and switch over our cars.

❺ Our enslavement to black gold goes much further than the problem of getting from Point A to Point B. We also need to keep the lights on. And this would be possible, for the first month or so, because only 1% of America's electricity is generated from oil—coal carries the largest burden, along with natural gas, nuclear and hydroelectric power.

❻ But brownouts and blackouts would soon begin. Sure, our electricity is generated mostly from coal, but how would the coal be extracted without those diesel-guzzling yellow trucks? How would it be hauled to the power plants? (Remember, our trains all run on diesel, too). Heating and cooling our homes would suddenly get a lot more complicated, and our televisions and laptops would be just a few more weeks away from shutting off forever.

❼ Forget even trying to get to work anymore; we now have another set of problems to solve, especially if it's winter and our houses are getting cold. Can we quickly put together some solar panels and batteries? A wind turbine? What do we have growing in the back yard that can burn? Environmentalists have been nudging us to insulate our homes and generate electricity from renewable resources for a while now; this might be the time to start paying attention.

❽ It gets much worse still, of course, because a world without oil would quickly become a world without all of the products made from petroleum that we have come to know, love, and depend upon. The list of essentials that we'd soon be doing without is prodigious: virtually all plastics, paints, medicines, hospital machines that go "bleep", Barbie dolls, ballpoint pens, breast implants, golf balls...

❾ Eating would get tougher, too. If no one can truck in fresh veggies from across the country, we might be inclined to go back to basics and grow our own food. Local farmers would become a necessity, not just people who sell us honey at the street fair. That said, make sure to keep the food coming, fresh and fast, because it's going to be awfully difficult to refrigerate. Fishing might work, so you'd need to get a new rod while supplies last. Alas, most of them are made of plastic. Then again, so is fishing line.

⑩ It's an interesting thought experiment to picture a world suddenly without oil. Taken to its logical conclusion, it encompasses so much more: a complete and rapid breakdown of society, leading to desperation, lawlessness, wars, and untold suffering. The scenario is unreal, of course, because we could never shut off our oil supply in a day, and in any case, there are trillions of barrels of the stuff still in the ground, right?

⑪ Yet, in a simpler sense, it's not so unrealistic, because even if it will happen more gradually than laid out here, we will indeed run out of oil. Output has already peaked in the majority of countries and has been declining in the United States since 1971. A handful of countries are still increasing production, but not enough to offset even bigger declines elsewhere. There is lots of oil still in the ground (we've used about half of the planet's generous endowment), but while the end of oil may be many decades away, the beginning of the end is now.

⑫ It's not just at the drip of the final drop that the oil crisis begins. It is when production stagnates and begins its inexorable fall. That perilous moment, alas, is now. Our oil supplies are about to begin to fail us. As oil becomes more scarce, we have to get serious about finding new solutions to power our world.

⑬ We have time to plan—but not that much time. And so far, we've done very little to prepare for a world without oil.

Activity Four Plan Your Own Oil-Free World

What might result from the oil deficiency according to Reading B? Summarize the key points and share with your team members. Illustrate your own perception of the world without oil and write down your own solutions or plans to tackle a world without oil.

____

____

____

____

After-Class Activities

Activity One Enhance Your Perception

Following your own "doomsday scenario", get an in-depth understanding of the "aftermath" by Reading C and summarize the supposedly affected areas the writer probes into in time order.

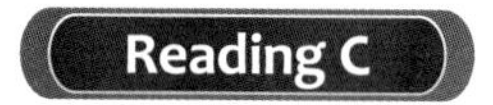
Reading C

A World Without Oil: The Aftermath[1]

James Kenny

Introduction

❶ Oil, is quite simply the backbone of modern life. It's in the food we eat, the houses we live in, and in the cars we drive. It's probably the most important commodity in the world; indeed this cheap, economic fuel makes our modern world possible. Over the last 150 years, we have taken about a trillion barrels from the Earth, and most experts forecast that the equivalent of another trillion should still be there for us to extract.

❷ But let's conduct a little thought experiment; most of us are aware that we are living on borrowed time. We know that the oil will run dry someday in the future—we're not entirely certain when, but most experts are confident that we will have to find an effective alternative fuel within the next 100 years or else. In our thought experiment, I want you to imagine that all of the remaining untapped oil reserves still in the Earth suddenly vanished overnight. What we would do? How would we cope with it?

❸ The experiment begins; the oil disappears and almost immediately oil refineries around the world go into chaos mode; alarms go off deep underground, indicating a major problem.

From Prosperity to Poverty

24 Hours Later

❹ News reports around the world confirm beyond doubt that all of the oil reserves below ground across the globe have disappeared. The oil companies move quickly to stem the rising

1. Kenny, J. (2016, June 24). *A world without oil: The aftermath*. Retrieved from https://soapboxie.com/social-issues/A-World-Without-Oil-Part-One

panic by informing people that there are 20 million barrels left in the refineries. Across the oceans, huge tankers carrying millions of barrels of oil are on the move, but not in the usual direction. In the wake of the crisis, Earth's biggest oil exporters such as Russia and Saudi Arabia have recalled their boats.

❺ This is a massive blow for the U.S., who is the biggest oil importer in the world. Each day, they produce more than 8 million barrels, but they actually consume double that amount. Now with the loss of imported oil, the deficit stands at 8 million and begins to grow.

❻ People quickly digest the news and flood into gas/petrol stations in the hope that they can fill up their cars for the last time, but many reports state that anyone preparing to queue faces a wait of two hours at least. Very quickly, gas/petrol stations around the world run dry; in the last frantic moments, many hike up the prices to truly astronomical amounts, and some of those determined people who sat in the queue hand over vast sums of money.

❼ Many countries around the world do have vast reserves of oil hidden away to deal with emergency situations similar to the one occurring now. The U.S. has roughly 725 million barrels of crude oil hidden away in secret location across the country. In order to protect what's left, the government takes dramatic steps, only allowing the most vital transports such as ambulances and fire trucks access to the oil. The age of planes, trains, and ships comes to a shuddering halt; roads become quiet, tracks empty, the skies quieter and cleaner. Each day in the U.S. alone, roughly 4 million people use aircraft for travel, but are now all stranded, forced to find alternative ways to get their destination. The loss of planes, trains, and ships also spells disaster for the delivery of cargo, over 100,000 tones of cargo will lie stranded, probably never to be delivered.

❽ The economic fallout is rapid; the growing, widespread panic forces the government to halt stock trading. The U.S. government took similar steps after the 9/11 disaster due to the panic that erupted in its aftermath. All of a sudden, two trillion dollars of oil stock become worthless; more than 400,000 people directly employed by the oil industry lose their jobs, and are reduced to have to find their way home by any means necessary. The uncertain economic future also forces thousands of manufacturing plants to shut down immediately, which spark protests from the millions employed in the industry who also lose their jobs.

❾ For the most part, we are largely ignorant of oil and just how important it really is. It's one of the most powerful and versatile fuels on the planet, made from dead organic matter that has been slowly compressed and heated over millions of years. It's in everything, from toothpaste, lipstick, polyester, and even plastic, but now it's all gone, and what was kept in reserve is dwindling rapidly. A huge chain reaction has now been set in motion, that is quickly crippling every part of our lives from hospitals, food, and of course power. The crisis is only just beginning.

Five Days Later

⑩ In just five short days, the loss of oil has forced governments around the world to declare martial law to stem the rising anger and anarchy among the population. In the U.S., the National Guard is deployed widely across the country, patrolling the streets of Los Angeles and Washington vigorously. The stock markets remain firmly shut, and unemployment has risen swiftly up to an astonishing 30%.

⑪ In less than a week, many of our most basic needs are suddenly out of reach. Food depot centres across the western world are now closed, sparking a major food crisis. Prior to the crisis, California for example sent out 1,300 trucks from its depots every day, delivering fresh food all over the country to grocery outlets. Now the trucks sit idle without their precious oil.

⑫ All of the big cities are hit hard; on average it takes one football field of farmland to produce enough food for just one person a year. Oil enabled the easy distribution of food from far and wide, but now without it, feeding a city of millions like New York becomes impossible. The trip to the grocery store now takes hours rather than minutes, and each outlet has a team of armed guards manning the doors, deciding how many people can enter the store at once. Inside the store, almost all of the best quality food has now gone, what's left are the ones that don't normally make it to the shelf, the ones that carry imperfections or are slightly off. But people can no longer afford to be fussy, and must make do with whatever they can find. The prices of food just like gas/petrol skyrocket, for example, in this new world a 5-lb bag of apples now costs nearly 12 dollars.

⑬ Roughly a quarter of all food consumed in the U.S. is imported from elsewhere, and with no more ships bringing fresh supplies, the food stocks dwindle dramatically. On farms, the loss of oil is even more dramatic, over the last fifty years farming has become industrial, with many containing hundreds if not thousands of cows and other livestock. On average, a cow needs around 100 lbs of food a day, while a pig needs around 8 lbs. In a bitter ironic twist, these animals raised to feed humans face starvation themselves.

⑭ The loss of oil causes power systems around the world to fail, plunging the world into darkness. Around the world, roughly 40% of electricity comes directly from coal burning power plants. In the U.S., Florida is hit hardest, as it mainly relies on electricity generated directly from burning oil. The major hospitals in cities such as Miami and Orlando are equipped with emergency backup generators, but even these rely heavily on diesel fuel processed from oil. In San Francisco, law and order breaks out. In the middle of the night, looters emerge en masse. But as well as looking for food, they search for cooking oil that can be converted into fuel for diesel engine cars.

One Month Later

⑮ Governments around the world initiate a global shutdown, keeping only the most essential services operational. The emergency oil reserves are converted into diesel fuel for cargo trains that deliver coal to power plants, in an effort to restore power. The strategy works, and some basic electrical services are restored, but only in certain areas, as the electrical grids are no longer interconnected.

⑯ Florida is still in a state of blackout; the emergency fuel gets the trains running again, but instead of people, they carry food. The U.S.' oil supply continues to dwindle; even the most optimistic forecasts estimate that the U.S. only has 11 months worth of oil left. The gasoline or petrol powered car becomes obsolete; this is a total disaster for the U.S., as it is specifically built to serve cars. More than half of the population live in sprawling suburbs, and prior to the crisis had most of their food delivered straight to huge grocery outlets nearby. For the average American, the easy life has vanished; alternative measures must be found.

⑰ Out in the Midwest, farmers begin planting new crops to replace the usual fruit and vegetables. They select soya beans that contain oil which can be extracted and turned into diesel fuel. Corn is another crop that contains a fuel alternative and is grown extensively across millions of acres of land. The treasure yielded by this crop is ethanol which can be used to power gas/petrol powered cars.

⑱ While the U.S. roads sit empty, the story in Rio de Janeiro, Brazil is quite different. The roads are still packed with cars that are powered by ethanol extracted from sugar cane. In terms of biofuel production, the Brazilians are decades ahead of the Americans and other western nations. However, hope remains eternal that the end of the modern world hasn't arrived. Thousands of electric cars are still on the road and could pave the way for a better future. But back in the present, a more immediate and concerning challenge looms: the onset of winter in the northern hemisphere.

Five Months Later

⑲ The U.S. government announces the takeover of three of the biggest car manufacturers in the country. They intend to concentrate on producing electric trucks to help supplement the much needed food deliveries. Across the vast agricultural lands of the Earth, farmers take inspiration from Brazil and start planting sugar cane to speed up the production of ethanol.

⑳ However, in big cities across the U.S. and indeed the rest of the world, food terminals begin to close, resulting in a fast spreading famine. People form crushing queues at train stations waiting for food deliveries. Instead of fresh produce, they must make do with powdered milk and rice. The essential services such as coal delivery and emergencies are still operational; surviving on the ever dwindling oil reserves, everything else is at a standstill. The U.S. continues to dramatically reduce its oil consumption in order to stretch out the vital remaining reserves. But in just a few short months, there won't be enough oil for any food deliveries at all.

㉑ While food continues to be brought in, rubbish/garbage is no longer collected and taken away. In fact, any rubbish clear-up is a luxury afforded only to a few lucky people. The situation

sounds bad in the North, but things are even worse in Saudi Arabia, which is in the midst of an economic disaster. 90% of their income from exports came from oil; with all that gone, the country collapses into ruin. Japan was one of the biggest importers in the world; roughly 60% of all its nutritional needs came from overseas. With no ships docking at their ports, the entire population face starvation.

㉒ Back in the U.S., many people are no longer prepared to wait for the government to find a solution. Instead, they take matters into their own hands and start converting garages and basements into makeshift laboratories, where they conduct experiments in producing their own biofuel using scavenged chemicals such as methanol. This is a very dangerous experiment, but if it works, it could provide escape from starving cities. However, this ingenuity only works on diesel cars and for a limited time.

㉓ Other alternative fuel sources also face similar hurdles; while the soya bean harvest was more than the double what it was in the previous year, it only produced half a billion gallons of biofuel which is less than 1% of the diesel North America used each year prior to the crisis. Furthermore, no more can be produced until the next harvest. Farmers continue to valiantly plant more corn to gain a higher yield of ethanol.

㉔ A world without oil forces governments around the world to make tough and brutal decisions. In this case, should they tell the farmers to plant crops for food or fuel? Hospitals are rapidly running out of supplies; everything from rubber gloves, gowns, medicine, and lubricants all need oil in their manufacture. Without such necessaries, drug resistant infections become rampant. In the big cities, families are surviving literally by the skin of their teeth. But now matters take a turn for the worst. An electrical transformer fire which was mostly nothing more than a nuisance in a world with oil, becomes nightmarish. Abandoned vehicles block access, preventing emergency vehicles from dealing with the problem. The fire quickly spreads, before an explosion rocks the neighborhood.

㉕ Winter has arrived in the northern hemisphere, and for the millions of people living in northern cities, the time has come to make a tough decision. Do they lie low and wait for winter to pass? Or do they flee south? For many, there is no decision, as temperatures plummet, people flee en masse, heading south; in what is the biggest mass migration in human history. In the U.S., people from the north pour into the southern states and Mexico. While in Europe, people from Scandinavia, the Baltic, Russia, and Britain pour into southern France and the Mediterranean countries. The refugees travel on foot; cars are abandoned, including the ones that were revived by cooking oil. The cold has turned the fuel thick and sludgy; consequently the fuel lines and engines have become blocked, rendering the cars totally useless. The northern cities transform into eerie ghostly islands poking out of the snow and ice. Amazingly, not all people flee south; some hardy travelers venture north seeking out isolated cabins in the wilderness where they can sit out the winter living off a stockpile of food that they bought with them. However, the stockpile is limited to basically what they can carry on their backs. So they have to resort to hunting and trapping animals. On average, an adult male human needs around 210,000 calories to see him through a winter; thus the regular acquisition of fresh meat is a necessity for survival.

Activity Two Watch the Videos for More Perspectives

Watch the following two videos on the same topic for more descriptions from another perspective. Summarize the key points portrayed in these videos.

Video 1: World Without Oil—What if All the Oil Ran Out?

 **Video 2:** Welcome to a World Without Oil

Activity Three Maintain Your Own Stand

Do you believe a world without oil depicted in the articles above would happen in the foreseeable future? Why or why not? Search some references for data or facts and write a short essay with not more than 200 words to support your argument.

Integrated Exercises

I Read the following academic words, and check whether you can use them appropriately. For those you can not, look up in a dictionary or search online about their contextual use. Write down notes to strengthen your memory.

Reading A

1. commodity
2. vehicle
3. integral
4. component
5. antibiotics
6. contributor
7. emission
8. decline
9. concentration
10. dramatically
11. consecutive
12. renaissance

13. abundant
14. stable
15. mindset
16. peddle (*vt.*)
17. compatible
18. adequate
19. affordable
20. reliable
21. deplete
22. reside

Reading B

1. dismantle
2. stack
3. tow
4. overwhelming
5. virtually
6. confront
7. convert
8. delivery
9. infrastructure
10. enslavement
11. extract
12. haul
13. environmentalist
14. nudge
15. insulate
16. essential (*n.*)
17. prodigious
18. inclined
19. necessity
20. refrigerate
21. encompass
22. breakdown
23. desperation
24. scenario
25. peak
26. offset
27. endowment
28. handful
29. drip
30. stagnate
31. inexorable
32. perilous

Reading C

1. backbone
2. equivalent
3. chaos
4. mode
5. prosperity
6. refinery
7. tanker
8. recall
9. massive
10. deficit
11. digest
12. frantic
13. astronomical
14. vital
15. strand
16. destination

17. fallout
18. halt
19. erupt
20. versatile
21. compress
22. dwindle
23. cripple
24. anarchy
25. deploy
26. patrol
27. vigorously
28. unemployment
29. depot
30. spark
31. distribution
32. imperfection
33. fussy
34. emergency
35. livestock
36. starvation
37. looter
38. initiate
39. shutdown
40. cargo
41. optimistic
42. obsolete
43. specifically
44. sprawl
45. eternal
46. loom
47. onset
48. hemisphere
49. supplement
50. inspiration
51. terminal
52. famine
53. consumption
54. luxury
55. makeshift
56. ingenuity
57. hurdle
58. valiantly
59. brutal
60. lubricant
61. infection
62. rampant
63. literally
64. nuisance
65. nightmare
66. transformer
67. plummct
68. migration
69. flee
70. refugee
71. consequently
72. render
73. poke
74. hardy
75. wilderness
76. stockpile
77. resort
78. acquisition

II Decide on the contextual meaning of the following terms and expressions.

Reading A

1. building block:
2. synthetic fabrics:
3. heart valve:
4. life vest:
5. agriculture sector:
6. be fundamental to our modern way of life and high standard of living:
7. the total greenhouse gas emissions:
8. the oil and gas sector:
9. the stale mindset:
10. the environmental stewardship:
11. rise out of poverty:

Reading B

1. oil rigs:
2. the BP spill:
3. the Strategic Petroleum Reserve:
4. the overwhelming majority of the oil:
5. finished petroleum products:
6. aviation fuel:
7. gauge needles:
8. grocery stores/outlets:
9. establish delivery infrastructure:
10. nuclear and hydroelectric power:
11. brownouts and blackouts:

12. solar panels:

13. wind turbines:

14. renewable resources:

15. fresh veggies from across the country:

16. a breakdown of society:

17. untold suffering:

Reading C

1. run dry:

2. the remaining untapped oil reserves:

3. oil refinery:

4. stem the rising panic:

5. an alternative fuel:

6. in the wake of the crisis:

7. hike up the prices:

8. deal with emergency situations:

9. crude oil:

10. come to a shuddering halt:

11. spell disaster for:

12. lie stranded:

13. halt stock trading:

14. oil stock:

15. organic matter:

16. set in motion a huge chain reaction:

17. declare martial law:

18. stem the rising anger and anarchy:

19. basic needs:

20. food depot center:

21. spark a major food crisis:

22. prior to the crisis:______________________________
23. grocery outlets:______________________________
24. distribution of food:______________________________
25. fresh supplies:______________________________
26. food stocks:______________________________
27. coal burning power plants:______________________________
28. generate electricity:______________________________
29. emergency backup generators:______________________________
30. diesel fuel:______________________________
31. the electrical grids:______________________________
32. in a state of blackout:______________________________
33. sugar cane:______________________________
34. pave the way for a better future:______________________________
35. food delivery:______________________________
36. food terminals:______________________________
37. fresh produce:______________________________
38. the powered milk:______________________________
39. oil consumption:______________________________
40. stretch out the vital remaining reserves:______________________________
41. in the midst of an economic disaster:______________________________
42. collapse into ruin:______________________________
43. drug resistant infections:______________________________
44. electrical transformer:______________________________
45. emergency vehicles:______________________________
46. the regular acquisition of fresh meat:______________________________
47. abandoned vehicles:______________________________
48. lie low:______________________________
49. mass migration:______________________________
50. live off the stockpile of food:______________________________

51. see him through a winter:______________________________

Ⅲ Analyze the grammatical structure of the following complex sentences, figure out the meaning of each sentence, and paraphrase them.

1. The men and women of the oil and gas industry reject the stale mindset of last century's thinking peddled by some that oil and gas production and environmental stewardship are not compatible. (Reading A, Para. 7)

2. The list of essentials that we'd soon be doing without is prodigious: virtually all plastics, paints, medicines, hospital machines that go "bleep", Barbie dolls, ballpoint pens, breast implants, golf balls...(Reading B, Para. 8)

3. Taken to its logical conclusion, it encompasses so much more: a complete and rapid breakdown of society, leading to desperation, lawlessness, wars, and untold suffering. (Reading B, Para. 10)

4. People quickly digest the news and flood into gas/petrol stations in the hope that they can fill up their cars for the last time, but many reports state that anyone preparing to queue faces a wait of two hours at least. (Reading C, Para. 6)

5. Very quickly, gas/petrol stations around the world run dry; in the last frantic moments, many hike up the prices to truly astronomical amounts, and some of those determined people who sat in the queue hand over vast sums of money. (Reading C, Para. 6)

6. All of a sudden, two trillion dollars of oil stock become worthless; more than 400,000 people directly employed by the oil industry lose their jobs, and are reduced to have to find their way home by any means necessary. (Reading C, Para. 8)

7. Instead, they take matters into their own hands and start converting garages and basements into makeshift laboratories, where they conduct experiments in producing their own biofuel using scavenged chemicals such as methanol. (Reading C, Para. 22)

8. On average, an adult male human needs around 210,000 calories to see him through a winter; thus the regular acquisition of fresh meat is a necessity for survival. (Reading C, Para. 25)

Unit 2

Petroleum and Economy

Unit Objectives

Goal 1

Get well informed of what determines oil prices and why they often fluctuate through task-based activities.

Goal 2

Understand how oil prices impact economy by the case study of American economy.

Goal 3

Make an in-depth study of how to hedge against the oil price volatility or other oil trade risks.

The modern petroleum industry began in 1859 in Pennsylvania, when a person named Edwin L. Drake created the first oil well, a facility for extracting petroleum from natural deposits. Since then, petroleum has become a valuable commodity in industrialized parts of the globe, and oil companies actively explore for crude oil deposits and build large oil extraction facilities. With oil's stature as a high-demand global commodity comes the possibility that major fluctuations in price can have a significant economic impact.

Pre-Class Activities

Activity One Search for the Definitions

Search online for the definitions of the following terms or concepts and share your findings with your team members.

1. gasoline price:________________
2. crude oil price:________________
3. spot price:________________
4. Brent Blend:________________
5. West Texas Intermediate (WTI):________________
6. oil futures:________________
7. oil price volatility:________________

Activity Two Read the Price Chart

Can you read the prices listed in the following charts? Hold a team discussion and interpret them in turns.

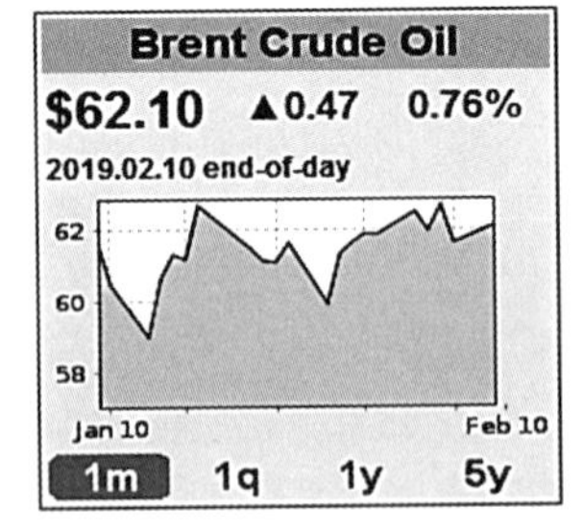

Crude oil	52.72	+0.15%
Natural gas	2.58	+1.24%
Gasoline	1.45	+1.42%
Heating oil	1.91	+0.42%
Gold	1313.70	+0.33%
Silver	15.77	+0.62%
Copper	2.81	-0.66%
2019.02.10 end-of-day	» Add to your site	

Figure 2-1 Crude oil and commodity prices

(Source: Oil-Price. Net.)

Activity Three Read the Price Components

Work in pairs and read the following figure to understand what constitutes the gasoline price.

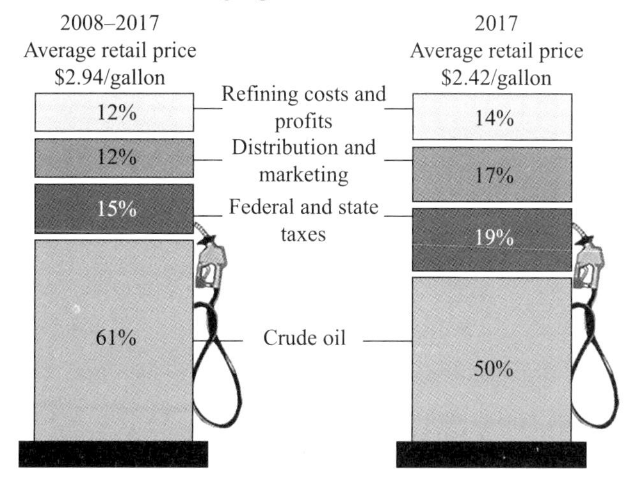

Figure 2-2 Percentage of what we pay for per gallon of retail regular grade gasoline

(Source: U.S. Energy Information Administration, Gasoline and Diese/Fuel Update.)

Activity Four Watch the Videos About Oil Price Determinants

Watch the following two videos to summarize the key points that determine the oil price and share your answer with your team members.

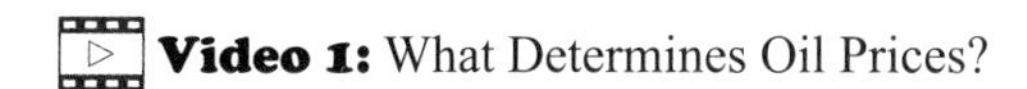

Video 1: What Determines Oil Prices?

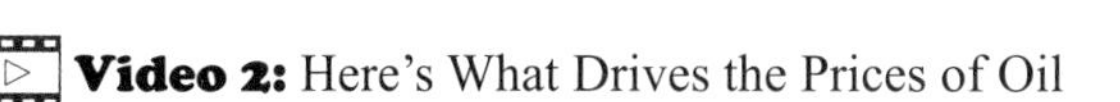

Video 2: Here's What Drives the Prices of Oil

While-Class Activities

Activity One Broaden Your Views

After watching the videos above, read the following article for more views on price determinants. Report orally what you've got through both the videos and the article to your team members .

Reading A

What Affects Oil Prices? Three Critical Factors[1]

Kimberly Amadeo

❶ Oil prices are controlled by traders who bid on oil futures contracts in the commodities market. That's why oil prices change daily. It all depends on how trading went that day.

❷ Other entities can only affect the traders' bidding decisions. These influencers include the U.S. government and the Organization of Petroleum Exporting Countries. They don't control the prices because traders actually set them in the markets.

❸ The oil futures contracts are agreements to buy or sell oil at a specific date in the future for an agreed-upon price. They are executed on the floor of a commodity exchange by traders who are registered with the Commodities Futures Trading Commission. Commodities have been traded for more than 100 years. The CFTC has regulated them since the 1920s.

❹ Commodities traders fall into two categories. Most are representatives of companies who actually use oil. They buy oil for delivery at a future date at the fixed price. That way, they know the price of the oil, can plan for it financially, and so reduce or hedge the risk to their corporations. Traders in the second category are actual speculators. Their only motive is to make money from changes in the price of oil.

Three Factors Traders Use to Determine Oil Prices

❺ There are three main factors that commodities traders look at when developing the bids that create oil prices.

❻ First is the current supply in terms of output. Since 1973, OPEC has a limited supply of 61% of the world's oil exports. But U.S. shale oil production doubled between 2011 and 2014. That created an oil glut. Traders bid the price down to $45 per barrel in 2014. Prices fell again in December 2015 to $36.87 a barrel. OPEC would normally cut supply to keep oil at its target of $70 a barrel. This time, it allowed prices to fall since it wouldn't lose money until oil is $20 a barrel.

❼ Shale producers need $40–$50 a barrel to pay the high-yield bonds they used for financing. OPEC bet that the shale oil producers would go out of business. This would allow it to keep

1. Amadeo, K. (2018, September 01). *What affects oil prices? Three critical factors*. Retrieved from https://www.thebalance.com/how-are-oil-prices-determined-3305650

its dominant market share. That started to occur in 2016. The oil price forecast has shown such volatility in prices because of the changes in oil supply, dollar value, OPEC's actions, and global demand.

❽ Second is access to future supply. That depends on oil reserves. It includes what's available in U.S. refineries as well as in the Strategic Petroleum Reserves. These reserves can be accessed very easily to increase oil supply if prices get too high. Saudi Arabia can also tap into its large reserve capacity.

❾ Third is oil demand, particularly from the United States. These estimates are provided monthly by the Energy Information Agency. Demand rises during the summer vacation driving season. To predict demand, forecasts for travel from AAA (American Automobile Association) are used to determine potential gasoline use. During the winter, weather forecasts are used to determine potential home heating oil use.

How World Crises Impact Oil Prices

❿ Potential world crises in oil-producing countries dramatically increase oil prices. That's because traders worry the crisis will limit supply.

⓫ That happened in January 2012 after inspectors found more proof that Iran was closer to build nuclear weapons capabilities. The United States and the European Union began financial sanctions. Iran threatened to close the Strait of Hormuz. The United States responded with a promise to reopen the Strait with military force if necessary. The possibility of an Israeli strike was also a concern.

⓬ As a result, oil prices bounced around $95 to $100 a barrel from November through January. In mid-February, oil broke above $100 a barrel and stayed there. Gas prices also went to $3.50 a gallon. Forecasts were that gas would be at least $4 a gallon through the summer driving season.

⓭ World unrest also caused high oil prices in the spring of 2011. In March 2011, investors became concerned about unrest in Libya, Egypt, and Tunisia in what became known as the Arab Spring. Oil prices rose above $100 a barrel in early March and reached its peak of $113 a barrel in late April.

⓮ The Arab Spring revolts lasted through the summer and resulted in an overturn of dictators in those countries. At first, commodities traders were worried that the Arab Spring would disrupt oil supplies. But when that didn't happen, the price of oil returned to below $100 a barrel by mid-June.

⓯ Oil prices also increased $10 a barrel in July 2006 when the Israel-Lebanon war raised fears of a potential threat of war with Iran. Oil rose from its target of $70 a barrel in May to a record-high of $77 a barrel by late July. A review of oil price history explains what makes oil prices so unpredictable.

Effect of Disasters on Oil Prices

⑯ Natural and man-made disasters can drive up oil prices if they are dramatic enough. Hurricane Katrina caused oil prices to rise $3 a barrel and gas prices to reach $5 a gallon in 2005. Katrina affected 19% of the nation's oil production. It came on the heels of Hurricane Rita. Between these two, 113 offshore oil and gas platforms were destroyed and 457 oil and gas pipelines were damaged.

⑰ In May 2011, the Mississippi River flooding caused gas prices to rise to $3.98 a gallon. Traders were concerned the flooding would damage oil refineries.

⑱ On the other hand, the Exxon-Valdez oil spill did not cause oil prices to rise. One reason was that oil prices in 1989 were only around $20 a barrel. The other was that only 250,000 barrels were spilled. Although this had a devastating impact on the Alaskan coastline, it didn't really threaten world supply.

⑲ The BP oil spill spewed more than 18 times the oil than did the Exxon Valdez. Yet, oil and gas prices barely budged as a result. Why? For one thing, global demand was down thanks to a slow recovery from the 2008 financial crisis and recession. Second, even though 174 million gallons of oil was spilled, it was over a long period of time. It also wasn't a large percentage of total oil used by the United States. In fact, it was only about nine days worth of oil. The United States consumed 6.99 billion barrels in 2010, according to the U.S. Energy Information Administration.

⑳ That's a little over 19 million barrels per day.

Activity Two Analyze the Difficulty in Forecasting Oil Prices

As discussed above, traders are likely to bid on oil futures contracts for oil prices. Are oil futures dependable? Bill Gilmer, director of the Institute for Regional Forecasting at the University of Houston's Bauer College of Business remarks: "In principle, it should be a very good predictor. But in fact, using the futures price as a forecast of the spot price of oil is a very small improvement over predicting that oil prices will be the same tomorrow as they are today." Read the article below for his standpoints and summarize the key points why futures market predictions can't do better.

Reading B

Why Are Oil Prices So Hard to Forecast?[1]

Bill Gilmer

Director of the Institute for Regional Forecasting

Oil Futures as a Spot Price Forecast

❶ This latest forecasting led me to the crude oil futures market, an often-quoted and much-maligned forecast of oil prices. In principle, it should be a very good predictor. But in fact, using the futures price as a forecast of the spot price of oil is a very small improvement over predicting that oil prices will be the same tomorrow as they are today. That sounds terrible, until you learn that futures market predictions beat all the alternatives, including other financial models, statistical models, and expert surveys. Why can't we do better?

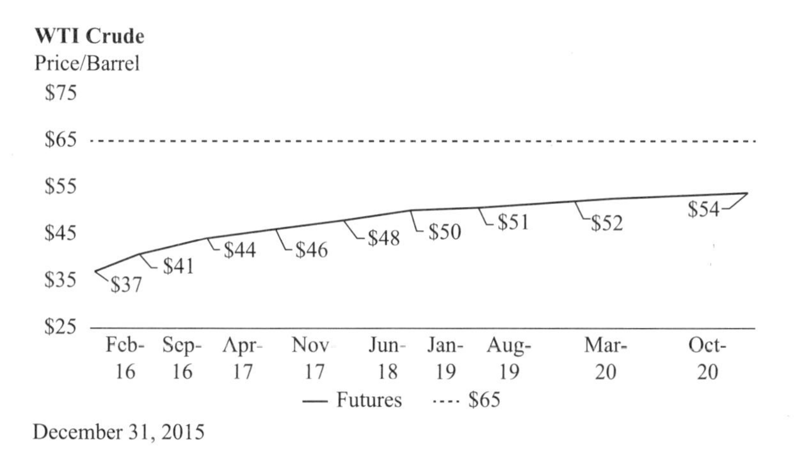

Figure 2-3 Tendency of WTI oil price implied by the futures market

❷ Figure 2-3 shows prices on the futures strip for NYMEX crude oil on December 31, 2015. At each date, the price is the payment that would be made and received for a barrel of West Texas Intermediate (WTI) delivered at that time. In the 1930s, it was thought that the spot or current price and all futures prices were independent, each determined by economic fundamentals prevailing at

1. Gilmer, B. (2016, January 19). Why are oil prices so hard to forecast? *Forbes*. Retrieved from https://www.forbes.com/sites/uhenergy/2016/01/19/why-are-oil-prices-so-hard-to-forecast/#28829d392740

that point in time.

❸ In the 1940s, agricultural economist Holbrook Working showed that spot and futures prices were closely linked by the cost of storage. If the 12-month futures price was higher than the spot price plus the cost of 12 months of storage, for example, I should buy inventory today, store it and sell it at a profit later. By the 1970s, economists had worked out how producers, consumers, hedgers, and speculators take a history of past prices, inventories, and market fundamentals, arbitrage across time, and the market simultaneously solves for the spot price and futures prices. It also turns out futures prices can be regarded as a forecast of oil prices. For example, the December 2017, futures price in Figure 2-3 is $48 per barrel, implying that will be the spot price on that date. If you live in Houston, this is a very gloomy outlook. We probably need $65 per barrel to put the fracking industry back to work, and perhaps allow it to grow moderately. Futures don't see a price near that level before 2020. How seriously do we take this forecast?

Crude Oil Futures

❹ Futures markets for grains and cotton were in full swing by the 1870s, but exchanges for crude oil and other energy products weren't established for another century. Heating oil was the first NYMEX energy product in 1978, followed by WTI crude in 1983 and later by gasoline and natural gas. The delay for crude and oil products was because much of the world's oil changed hands at posted or official prices until 1986, with the prices negotiated between large national oil producers and major oil companies. The demise of this system allowed today's futures market for crude to grow and rival the largest exchanges in the world, including commodities such as corn and copper.

❺ Early studies of crude oil futures as a forecast of spot prices were deeply divided. From one study to another, the markets were/were not efficient, or futures prices were/were not good forecasters. Many of these studies were premature, as it takes years to accumulate the data needed for good studies. To get around the lack of data, early studies too often relied on prices from the fixed-price regime of the 1970s and 1980s.

❻ To see what we know about these markets today, I found four relatively recent studies of crude oil and oil product markets; none of them used data from before 1990. This brief summary sounds very much like Tomek's conclusions for agricultural products.

❼ Crude oil markets are probably weak-form efficient. Three of the four studies support the notion across all the futures horizons studied.

❽ The studies typically show that the futures price forecast can beat a random walk, i.e., it is better than a naive forecast that says tomorrow will be the same as today.

❾ But futures are rarely better than a random walk by statistically significant margins. We can't be 90% or 95% sure futures are better.

⑩ Both futures prices and a random walk predict spot prices better than other financial or statistical models. For example, the study from the IMF looked at two alternative financial models and six alternative time-series models. Once more, futures beat out the random walk by a small margin, but the accuracy of other models fell far short of either futures or a random walk.

Why Oil Prices Are Hard to Predict

⑪ We have dug ourselves into a pretty deep hole. Futures prices are a poor predictor of spot prices, barely beating a random walk, but standard statistical models are even worse. Since futures markets are weak-form efficient, no financial model, statistical technique, or subjective survey based on public data should do better.

⑫ Why are all the forecasts so poor? It is because the world will not stand still. All of the evaluations of crude futures markets assume that on a particular day the market takes past prices, inventory data, and other fundamentals to produce a set of spot and futures prices. We write down the 12-month futures price, for example, then wait a year and check the spot market to see if the forecast was right.

⑬ But that forecast was completely predicated on information available a year ago. We can all think of moments that have suddenly and unexpectedly turned oil markets on their head: the Arab oil embargo, the fall of the Shah, or the invasion of Kuwait. An efficient market scrapes together all available data and uses it to look forward, but no one should pretend it can somehow divine the future.

⑭ And it doesn't take big headlines to upset the forecast. The global crude oil market depends on the politics of dozens of producing countries, economic cycles in consumer countries, and a vast infrastructure of pipes, ships, and refineries. Even if we account for the known issues correctly, we could list 1,000 or more low-probability events that could push our forecast off course.

⑮ Suppose that each of these events has a probability of one in a 1,000 over the next 12 months. There is no reason to incorporate any of these possibilities into our forecast or even list them as a risk. But if these events are independent of each other, the chance that at least one will significantly and unexpectedly affect the oil market within a year is $1-(.999)^{1000}$ or 63.2%.

⑯ When I opened the newspaper December 31 and looked at the futures prices in Figure 2-3, what was I reading? Was the 12-month futures contract at $44 telling me what the spot price of crude oil will be a year from now? Probably not very accurately, because it is not clairvoyant; unanticipated events in crude markets over the next 12 months—those constantly changing facts—leave the futures price barely more capable than a random walk.

⑰ When important new information changes the December 31 outlook, has the futures forecast failed? No, the world changed and the futures market quickly updated its forecast to include new data—efficiently, as far as we know. As long as the world does not stand still, neither will the

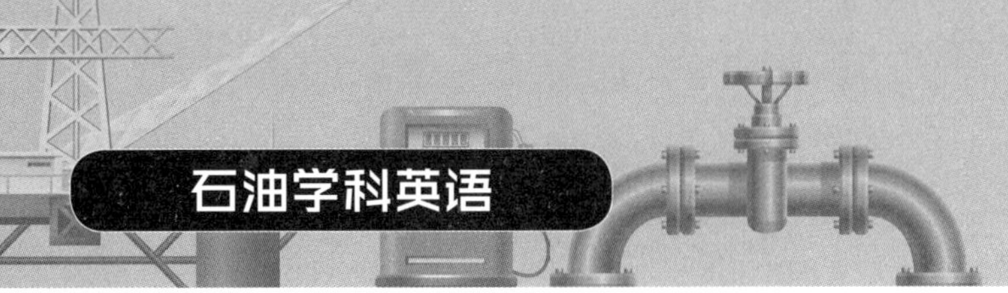

futures price.

⑱ But if on December 31, you wanted the best oil price forecast possible based on the facts available that day, you wanted the crude futures prices. The forecast is available daily, updated continuously and all for the price of a newspaper.

Activity Three Understand the Impact of Oil Prices on Economy

Oil prices are regarded as a barometer to economy. Read the following article to comprehend how oil prices impact economy by the case study of American economy. Report orally your findings to the whole class about the key areas affected by oil prices.

How Oil Prices Impact the U.S. Economy[1]

Andrew Beattie

❶ The extraction of oil and natural gas from shale has reduced the amount of oil the United States needs to import and is adding to the economy in the forms of jobs, investment, and growth. Oil exploration and production is again an important industry in the United States. In this article, we will look at how oil prices impact the U.S. economy.

A Reversal of Fortune

❷ In the 1990s and early 2000s, the United States was struggling under declining domestic oil production and the resulting need to import more oil. Wells in Texas and other regions were still producing, but falling far short of meeting growing energy demands. In the latter half of the 2000s, however, new technology allowed companies to economically draw oil and gas from shale deposits that were once considered depleted because the cost of extraction would be impractical.

❸ Higher prices per barrel of oil also helped to justify the cost of a hydraulically fractured well. The United States is once again one of the top producers of oil and gas. Greater domestic oil production is a net positive for the United States. However, as an oil-producing country (and not just an oil consumer), the United States now also feels an unpleasant pinch when oil prices drop.

Oil and the Costs of Doing Business

❹ The price of oil influences the costs of other production and manufacturing across the

1. Beattie, A. (2018, October 13). *How oil prices impact the U.S. economy*. Retrieved from https://www.investopedia.com/articles/investing/032515/how-oil-prices-impact-us-economy.asp

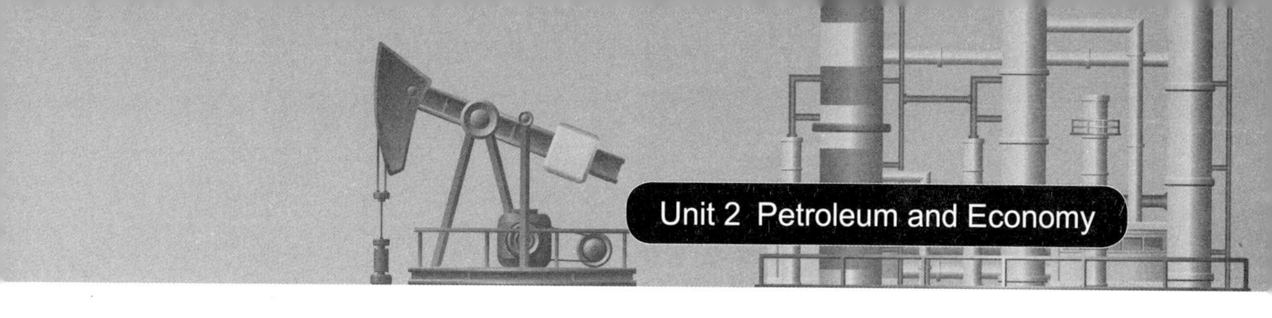

United States. For example, there is the direct correlation between the cost of gasoline or airplane fuel to the price of transporting goods and people. A drop in fuel prices means lower transport costs and cheaper airline tickets. As many industrial chemicals are refined from oil, lower oil prices benefit the manufacturing sector. Before the resurgence in U.S. oil production, drops in the price of oil were largely viewed as positive because it lowered the price of importing oil and reduced costs for the manufacturing and transport sectors. This reduction of costs could be passed on to the consumer. Greater discretionary income for consumer spending can further stimulate the economy. However, now that the United States has increased oil production, low oil prices can hurt U.S. oil companies and affect domestic oil industry workers.

❺ Conversely, high oil prices add to the costs of doing business. And these costs are areas also ultimately passed on to customers and businesses. Whether it is higher cab fares, more expensive airline tickets, the cost of apples shipped from California, or new furniture shipped from China, high oil prices can result in higher prices for seemingly unrelated products and services.

Job Growth and Investment Dollars

❻ The exploration and production of U.S. shale deposits have been a strong source of job growth. The hydraulically fractured wells tend to have a shorter production life, so there is always new drilling activity to find the next deposit. All this activity requires labor including drilling crews, loader operators, truck drivers, diesel mechanics, and so on. The people working in these areas then support surrounding businesses like hotels, restaurants, and car dealerships. Lower oil prices mean less drilling and exploration activity because most of the new oil driving the economic activity is unconventional and has a higher cost per barrel than a conventional source of oil. Less activity can lead to layoffs which can hurt the local businesses that catered to these workers. Therefore, the negative impact will be felt keenly in the shale regions even as some of the positive impacts of lower oil prices start to show in other regions of the United States. This is regionally painful for the country and effects show in state-level unemployment statistics.

❼ However, these losses may not have a noticeable impact on national unemployment numbers.

❽ The other groups that tend to suffer when U.S. oil prices drop are the banking and investment sectors. There are a lot of different companies drilling and servicing wells on the shale deposits, and many of these companies finance their operations by raising capital and taking on debt. This means that investors and banks both have money to lose if the price of oil drops to where new wells are no longer profitable and the companies dependent on drilling and service then go out of business. Of course, investors and bankers are well versed in risks and rewards, but the losses still destroy capital when they happen. Between the job losses and the capital losses, a dip in oil prices can trim the growth of the U.S. economy.

The Benefits of Diversity

❾ Even with the loss of growth, the U.S. economy isn't nearly as tied to the price of oil as some of the other top production nations. The U.S. economy is incredibly diverse. Although oil and gas production has been one driver of recent growth, it is far from the most important sector of the economy. It is, of course, connected to other sectors and losing growth in one can weaken others, but sectors like manufacturing gain more than they lose.

❿ The U.S. economy can take a lot of hits and keep on going because so many sectors contribute to it without any single dominant sector. The same can't be said about some other oil-producing nations like Russia or Venezuela whose fortunes rise and sink with the price of oil. In short, the U.S. economy has the room to adapt to prolonged periods of high or low oil prices. This means it takes more than just low oil to shake the U.S. economy, but it is not uncommon for oil prices, high or low, to increase the impact of economic shocks.

Bottom Line

⓫ Oil prices do have an impact on the U.S. economy, but it goes two ways because of the diversity of industries. High oil prices can drive job creation and investment as it becomes economically viable for oil companies to exploit higher-cost shale oil deposits. However, high oil prices also hit business and consumers with higher transportation and manufacturing costs. Lower oil prices hurt the unconventional oil activity, but benefit manufacturing and other sectors where fuel costs are a primary concern.

After-Class Activities

Activity One Get to Know Literature Reviews Cooperatively

Read the following excerpted literature review. First look up the references in bold to get their meanings and then hold a discussion with your team members about the purposes, chief features of writing a literature review. You are asked to make a class presentation to summarize your perception of what literature reviews are like and the key ideas of this selected review.

(The following is the first introduction paragraph of an article[1] *by S.Y. Kan, et al., part of the literature review session.)*

With the economic development and population growth, global demand for energy is projected to increase by 30% in 2035 **(BP, 2017b)**, putting an enormous burden on energy

1. Kan, S.Y., et al. (2019). Natural gas overview for world economy: From primary supply to final demand via global supply chains. *Energy Policy, 124*, 215-225. DOI: 10.1016.2018.10.002

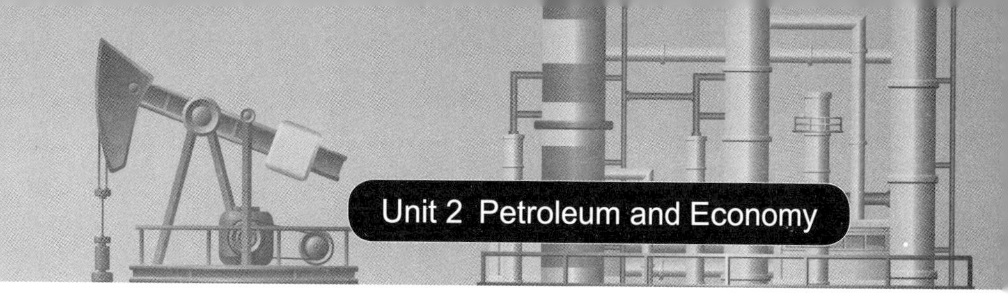

supply. There is also a worldwide concern about climate change and environment degradation, which can be largely attributed to the current carbon-intensive energy system **(IEA, 2018; IPCC, 2015)**. In this context, world economies are shifting towards a low-carbon energy mix **(IEA, 2017b)**. Natural gas is regarded as the cushion to smooth the transition towards energy systems dominated by renewable energy because it produces far less carbon emissions and pollutants than coal and oil **(Heath et al., 2014; Laurenzi and Jersey, 2013; Sun and Li, 2015)**. For example, it can help ICEs-buses reduce more than 30% CO_2 emissions in the combustion in China **(Sun and Li, 2015)**. Moreover, shale gas revolution has contributed significantly to reducing the cost of natural gas production **(Mistré et al., 2018)**. Therefore, global supply and demand of natural gas remain an upward trend, occupying an increasing share of fuel mix in more and more countries and regions **(BP, 2017b; IEA, 2017a)**. For instance, by 2040, natural gas is anticipated to account for 40% of total energy consumption in U.S.A. with its production rising by 65%, overtaking oil as the leading fuel **(BP, 2018b)**. In Russia, it is estimated to dominate in energy mix with a share of 50% and its production will rise up to 72 Bcf/d **(BP, 2018a)**. It can be expected that natural gas will play an ever-increasing important role in global economy. Due to the geographical separation between supply and demand of natural gas, significant amount of natural gas has entered global commodity market. Though natural gas pipelines are restricted by geopolitics, its market is globalizing thanks to LNG (liquefied natural gas) **(EIA, 2017; IGU, 2017)**. According to BP **(BP, 2017a)**, trade of natural gas in 2016 amounted to 1084 bcm (billion cubic meter), accounting for 31% of global natural gas production. In light of this great prospect, extensive researches have been carried out to understand the future of natural gas supply and demand **(Lu et al., 2016; Paltsev et al., 2011; Shaikh and Ji, 2016)**, international trade of natural gas **(Arora and Cai, 2014; Geng et al., 2014; Zhang et al., 2018)**, natural gas security **(Bouwmeester and Oosterhaven, 2017; Egging and Holz, 2016; Ruble, 2017)**, and so forth.

Activity Two Practice Summary Writing

Summary writing is viewed as an essential step towards writing a good literature review. The following is the interviewee expert Michael A. Levi's policy options on how to handle the oil price volatility as hedges against oil trade risks. Read it carefully and then write a summary about the key ideas the expert recommends with not more than 150 words.

Reading D

How to Handle Oil Price Volatility[1]

Michael A. Levi

Prices at the pump are emerging as a significant U.S. election issue. Five experts offer a range of policy options, from lowering regulations to encouraging less consumption. The author is an expert on American Council on Foreign Relations.

❶ There is a myth, popular among both politicians and the public, that high oil prices are the greatest economic risk that the United States faces when it comes to energy. They're wrong; wildly changing prices, not high ones per se, are what really do damage. Rapidly rising prices drain consumers' wallets without giving them time to adapt; frequent change also makes long-term investments more difficult. People may applaud when prices crash, but to turn a cliché on its head, what goes down must go up.

❷ Policymakers should focus their responses on two dimensions: steps that blunt intolerable volatility and ones that help consumers cope with the consequences of whatever remains.

❸ Some volatility is natural and quite tolerable. Markets aren't perfect predictors of the future, which means that prices will shift to and fro. Since there's no reason to think that governments would be smarter, they usually shouldn't try to override what the markets do. Moreover, modest volatility can prompt consumers to take steps, like shifting to more fuel-efficient cars that will help them if volatility later explodes.

❹ There are, however, exceptions to the general rule that the government should stay out of the market. Markets are ill-equipped to handle the sorts of large price swings that would result from major geopolitical events like, for example, a confrontation with Iran. Those sorts of occasions call for the government to use the Strategic Petroleum Reserve in order to buffer the market.

❺ Moreover, in many cases, other governments' market interference through things like oil subsidies makes volatility worse; there, U.S. diplomatic efforts to help reduce those distortions are wise.

❻ There is more, though, which the government can do to help consumers cope with. A fifty-dollar price swing is only half as bad if you're using half as much oil. Strict fuel economy regulations can steer people in that direction. So would a gas tax, perhaps as part of a comprehensive fiscal package, though the prospects remain remote. Helping consumers get access to hedging products—in essence, helping democratize oil derivatives rather than trying to shut them down—could also help them better deal with gyrations.

1. Adapted from an Expert Roundup interviewed by Toni Johnson in March 16, 2012.

❼ All of these would pay off over the long term. The best bet for the next few months is an increasingly wild ride.

Integrated Exercises

I Read the following academic words, and check whether you can use them appropriately. For those you can not, look up in a dictionary or search online about their contextual use. Write down notes to strengthen your memory.

Reading A

1. critical
2. entity
3. agreed-upon
4. register
5. regulate
6. category
7. financially
8. hedge
9. corporation
10. speculator
11. glut
12. barrel
13. dominant
14. forecast
15. volatility
16. available
17. estimate
18. predict
19. potential
20. inspector
21. capability
22. sanction
23. strike
24. bounced
25. unrest
26. revolt
27. disrupt (*vt.*)
28. record-high
29. unpredictable
30. dramatic
31. offshore
32. platform
33. gallon
34. devastate
35. recession
36. consume

Reading B

1. quote
2. malign
3. alternative
4. statistical
5. survey
6. strip

7. fundamental
8. prevailing
9. inventory
10. hedger
11. arbitrage
12. simultaneously
13. fracking
14. moderately
15. negotiate
16. demise
17. rival (*vt.*)
18. regime
19. naive
20. random
21. margin
22. technique
23. subjective
24. evaluation
25. assume
26. scrape
27. divine
28. probability
29. incorporate
30. accurately
31. swing
32. unanticipated
33. constantly
34. barely

Reading C

1. extraction
2. investment
3. reversal
4. declining
5. domestic
6. impractical
7. justify
8. pinch
9. correlation
10. resurgence
11. discretionary
12. stimulate
13. conversely
14. surrounding
15. unconventional
16. layoff
17. keenly
18. prolong
19. viable
20. exploit

Reading D

1. myth
2. applaud
3. cliché
4. blunt
5. intolerable
6. predictor
7. override
8. prompt
9. exception
10. buffer
11. interference
12. distortion

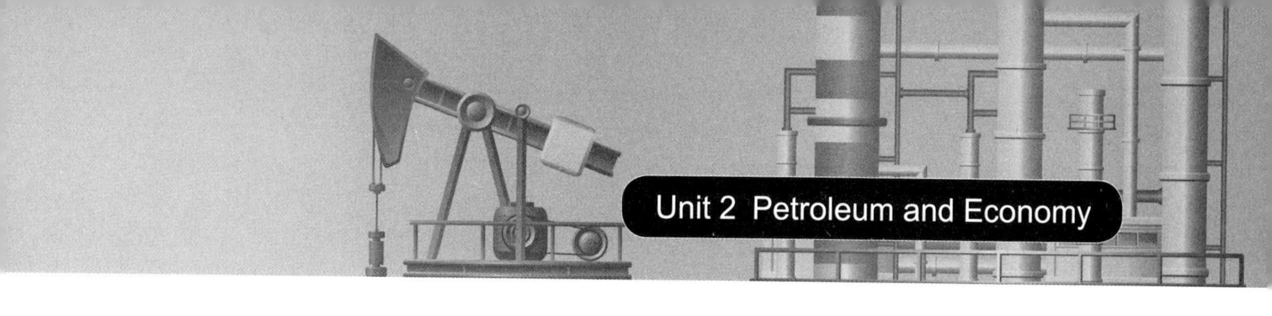

13. steer
14. fiscal
15. derivative

II Decide on the contextual meaning of the following terms and expressions.

Reading A

1. behind-the-scenes role:________________
2. bid on oil futures contracts:________________
3. commodities market:________________
4. the traders' bidding decisions:________________
5. fall into two categories:________________
6. shale oil:________________
7. cut supply:________________
8. high-yield bonds:________________
9. market share:________________
10. oil demand:________________
11. oil reserve:________________
12. reserve capacity:________________
13. nuclear weapons capabilities:________________
14. reach the oil price peak:________________
15. offshore oil and gas platforms:________________
16. financial sanctions:________________
17. financial crisis and recession:________________
18. commodities traders:________________

Reading B

1. the spot price of oil :________________
2. futures market predictions:________________
3. expert surveys:________________

4. financial models: ______

5. a gloomy outlook: ______

6. heating oil: ______

7. posted prices: ______

8. crude oil futures: ______

9. the fixed-price regime: ______

10. weak-form efficient: ______

11. a random walk: ______

12. by statistically significant margins: ______

13. standard statistical models: ______

14. subjective survey: ______

15. inventory data: ______

16. a set of spot and futures prices: ______

17. oil embargo: ______

18. economic cycles: ______

19. low-probability events: ______

20. push the forecast off course: ______

Reading C

1. oil exploration and production: ______

2. domestic oil production: ______

3. shale deposits: ______

4. the cost of extraction: ______

5. industrial chemicals: ______

6. manufacturing sector: ______

7. cab fares: ______

8. drilling crews: ______

9. diesel mechanics: ______

10. car dealerships: ______

11. loader operators:____________________

12. state-level unemployment statistics:____________________

13. the banking and investment sectors:____________________

14. capital losses:____________________

15. be well versed in risks and rewards:____________________

16. a dip in oil prices:____________________

17. trim the growth of economy:____________________

18. economic shocks:____________________

19. manufacturing sector:____________________

20. take a lot of hits:____________________

21. bottom line:____________________

22. a primary concern:____________________

Reading D

1. per se:____________________

2. shift to and fro:____________________

3. fuel-efficient cars:____________________

4. stay out of the market:____________________

5. price swings:____________________

6. oil subsidies:____________________

7. a comprehensive fiscal package:____________________

8. get access to hedging products:____________________

9. democratize oil derivatives:____________________

10. wild ride:____________________

III Analyze the grammatical structure of the following complex sentences, figure out the meaning of each sentence, and paraphrase them.

1. Yet, oil and gas prices barely budged as a result. Why? For one thing, global demand was down thanks to a slow recovery from the 2008 financial crisis and recession. (Reading A, Para. 19)

2. But in fact, using the futures price as a forecast of the spot price of oil is a very small improvement over predicting that oil prices will be the same tomorrow as they are today. That sounds terrible, until you learn that futures market predictions beat all the alternatives, including other financial models, statistical models, and expert surveys. (Reading B, Para. 1)

3. The delay for crude and oil products was because much of the world's oil changed hands at posted or official prices until 1986, with the prices negotiated between large national oil producers and major, oil companies. (Reading B, Para. 4)

4. Once more, futures beat out the random walk by a small margin, but the accuracy of other models fell far short of either futures or a random walk. (Reading B, Para. 10)

5. Probably not very accurately, because it is not clairvoyant; unanticipated events in crude markets over the next 12 months—those constantly changing facts—leave the futures price barely more capable than a random walk. (Reading B, Para. 16)

6. In the 1990s and early 2000s, the United States was struggling under declining domestic oil production and the resulting need to import more oil. Wells in Texas and other regions were still producing, but falling far short of meeting growing energy demands. (Reading C, Para. 2)

7. Therefore, the negative impact will be felt keenly in the shale regions even as some of the positive impacts of lower oil prices start to show in other regions of the United States. This is regionally painful for the country and effects show in state-level unemployment statistics. (Reading C, Para. 6)

8. This means that investors and banks both have money to lose if the price of oil drops to where new wells are no longer profitable and the companies dependent on drilling and service then go out of business. (Reading C, Para. 8)

9. High oil prices can drive job creation and investment as it becomes economically viable for oil companies to exploit higher-cost shale oil deposits. However, high oil prices also hit business and consumers with higher transportation and manufacturing costs. (Reading C, Para. 11)

10. Policymakers should focus their responses along two dimensions: steps that blunt intolerable volatility and ones that help consumers cope with the consequences of whatever remains. (Reading D, Para. 2)

Unit 3

Petroleum and Geopolitics

Unit Objectives

Goal 1

Get familiar with what petroleum politics is and the fundamentals of how petroleum affects politics through task-based activities.

Goal 2

Understand why geopolitics matters so much to the petroleum markets.

Goal 3

Make an in-depth study of why oil industry can exert great impact on global politics.

The politics of oil emerged in the 20th century as one of the most critical dimensions shaping domestic and global life. Little did the oil prospectors in Titusville, Pennsylvania, know in 1859 that they had struck upon a commodity that would prove central over the next 150-plus years in affecting issues of global poverty and economic growth, war and peace, terrorism, democracy, global power politics, global climate change, the rise of new great powers, and the decline of actors that used or pursued oil unintelligently. All of these dimensions constitute the politics of oil, a commodity that, perhaps like no other, has shaped global life and is likely to do so for the foreseeable future.

Pre-Class Activities

Activity One Search for the Definitions

Search online for the definitions of the following terms or concepts and share your findings with your team members.

1. oil diplomacy:____________________
2. geopolitics:____________________
3. petrodollar:____________________
4. oil embargo:____________________
5. OPEC:____________________
6. oil and international conflict:____________________
7. ten key oil producing countries:____________________
8. oil money:____________________

Activity Two Watch the Video About 1973 Oil Crisis and Its Impact

Watch the following video to summarize the key effects the oil crisis has on politics and share your answer with your team members.

 **Video:** Conflict in the Middle East: OPEC's 1970s Oil Embargo and Its Impact

While-Class Activities

Activity One Make a Seminar Discussion

The following is commonly quoted when it comes to the impact of petroleum on politics:

"Control of petroleum reserves has played an overriding role in international politics." Hold a discussion with your team members about geopolitics and present your viewpoints to the whole class.

Activity Two Get to Know Petroleum Politics

Based on your presentation, read the following article to further understand the underlying relationship between petroleum and politics. Summarize the key points on the close connections between oil and politics and report orally what you've learned to the whole class.

Reading A

Oil, Politics, and Power[1]

Claude Salhani

❶ Oil and politics have always gone together for a simple reason; since oil became an indispensable commodity without which the world as we know it today would not function, countries that produce oil have learned how to use it as a weapon. And who says weapons, says politics.

❷ The power of oil as a political weapon became evident during the 1973 Arab-Israeli conflict that became known as the October War in the Arab world and the Yom Kippur War in Israel. Hoping to sway Western sentiments in favor of the Arab causes Arab oil producing countries such as Saudi Arabia and the Gulf sheikdoms agreed to reduce their output. Naturally, less oil on the market meant higher prices at the pump and for the home consumer of heating oil. The Arab oil embargo forced Western governments to enact strict measures in order to safeguard oil reserves.

❸ The tactic employed by the oil producers however backfired: Forced by some governments to leave their cars in the garage on alternate days along with having to pay more money for less gas, the 1973 Arab oil embargo initiative was a public relations disaster. Furthermore, given that they were producing, exporting, and selling less gas, the oil producers lost billions of dollars in potential revenues.

❹ However, what the 1973 oil embargo did accomplish was to demonstrate the potential oil

1. Salhani, C. (2010, January 21). *Oil, politics, and power.* Retrieved from https://oilprice.com/Geopolitics/International/Oil-Politics-And-Power.html

had been as a weapon. The outcome changed much in the modern history of oil and politics. The embargo forced the West to become less dependent on Arab oil and American and international oil companies began looking elsewhere to supplement Arab oil.

❺ There were alternatives to Arab oil except that until the Arab embargo of 1973 purchasing Arab oil was far less expensive than erecting platforms in the inclement weather of the North Sea, for example, in Norwegian waters or off the English coast. The Arab oil embargo and rising oil prices justified exploitation of North Sea oil, Canadian oil, and other previously untapped oil fields. The outcome was two-fold: First, European and American dependence on Arab oil lessened; second, it gave the new producers additional revenues as oil prices continued to escalate.

❻ However, the importance of oil in politics, or rather the importance of the politics of oil, and the important role oil would play in modern post-World War Ⅱ geopolitics was recognized by the United States very early on. It soon became evident that in the industrialized era oil would replace coal as the main source of energy and as the coalmining towns of Newcastle and West Virginia began to die, a new mirage began to rise in the deserts of Arabia.

❼ The U.S.'s keen interest in oil politics surfaced around the close of WW Ⅱ, when on February 15, 1945, U.S. President Franklin D. Roosevelt flew to Egypt to meet with Saudi Arabia's King Abdulaziz ibn Saud aboard the USS Quincy in the Great Bitter Lake in the Suez Canal. The meeting between Roosevelt and inb Saud was a major landmark in contemporary history of oil as it opened the chapter of oil-politics when the American president promised the Saudi King to protect his oil fields in return for preferential treatment.

❽ Just weeks earlier, the United States had thwarted a final German attempt to make one last thrust through Allied lines at Bastogne, (where General Anthony McAullif is reported to have said "Nuts", when asked by the Germans to surrender). Had the Germans been successful, they would have been able to link their forces in the Ardennes with the Belgian port of Antwerp, thus giving them access to an uninterrupted flow of oil, essential to keep the gaz-guzzling tanks moving forward. As it turned out the Germans lost the Battle of the Bulge because their tanks ran out of gas.

❾ Roosevelt immediately recognized the strategic importance of oil. Had Nazi Germany won the Battle of the Bulge, World War Ⅱ would have been prolonged perhaps just long enough to allow German scientists to finalize the V-2 rocket and as they hoped, produce the first atomic bomb, giving them ultimate victory. In essence what lost the war for Germany was shortage of oil.

❿ Since the end of WWⅡ, there were other wars that were fought over oil. The United States went to war in 1990–1991 against Saddam Hussein to liberate tiny Kuwait from Iraq after Saddam's forces declared Kuwait was Iraq's 19th province and occupied it.

⓫ Indeed, one might even trace the events of 9/11 and Osama bin Laden's hatred of America for the nearly unconditional support given by the U.S. to the House of Saud to that historic meeting

in the Great Bitter Lake between Roosevelt and ibn Saud.

⓬ Finally, it is interesting to note that the two presidents who took America into wars in the Middle East over oil—President George Bush and his son, George W. Bush—both had connections to oil money. Coincidence? You decide.

Activity Three Embrace a New Perspective

There is a different line of thinking with regard to the current geopolitics of oil as discussed in the article below. What are the prominent changes of oil politics? Summarize the key points and share with your team members.

Reading B

Eight Reasons Why the Politics of Oil Have Changed[1]

Katinka Barysch
Group Strategy and Portfolio Manager, Allianz

❶ Energy is not an everyday commodity; it is highly political. The 70% oil-price fall since mid-2014 therefore raises a host of political questions.

❷ The international energy market has changed in fundamental ways:

❸ Firstly, oil is not a scarce resource any more. The geopolitical battle is no longer over access to resources but about global market share. In particular, Saudi Arabia seems intent on flooding global markets to push out higher-cost producers, especially in the U.S. But the oil glut could also have political reasons, such as undermining Russia. A more recent reason might be to prevent its regional rival Iran from re-entering the oil market, now that sanctions have been lifted.

❹ Iran, meanwhile, has little interest in cooperating with the Saudis on oil. That means that OPEC looks unlikely to be revived.

❺ Secondly. there is no longer a clear-cut front of oil producers vs consumers. Today, most of the big players are both. One-third of Saudi production is consumed at home. In 2014, the U.S. overtook Russia as the world's biggest energy (oil and gas) producer. This also makes coordinated action more difficult.

1. Barysch, K. (2016, February 19). *Eight reasons why the politics of oil have changed*. Retrieved from https://www.weforum.org/agenda/2016/02/eight-reasons-why-the-politics-of-oil-have-changed/

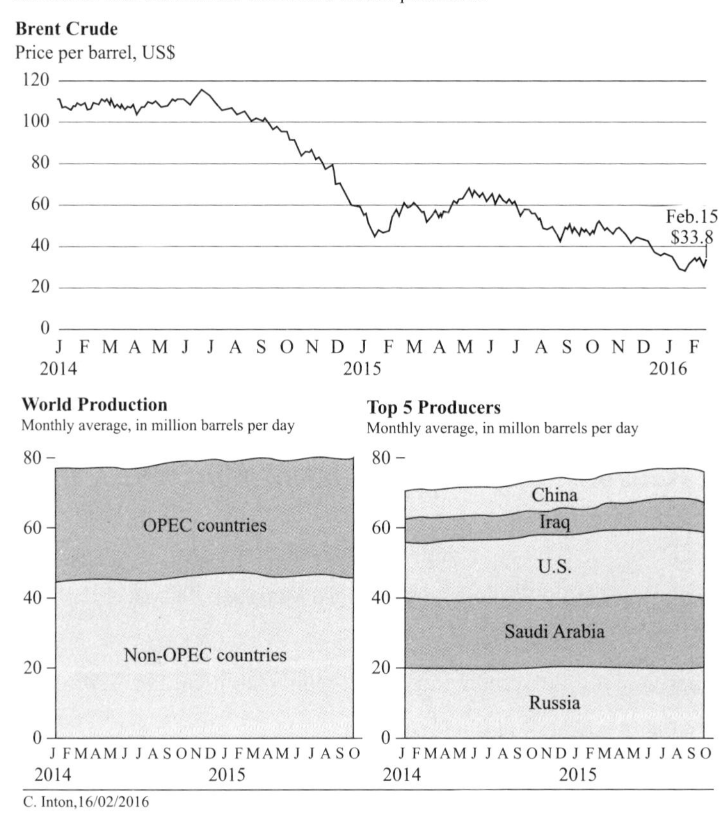

Figure 3-1　Crude oil production

(Source: Thomson Reuters: EIA. Data includes lease condensate.)

❻ Thirdly, climate policies are introducing new uncertainties on both the supply and demand side. If countries are serious about reaching a climate change target of 2 degrees Celsius (or even 1.5 degrees, as agreed in Paris), they simply cannot burn all the oil and gas that is still in the ground. Climate targets could leave a lot of energy resources as "stranded assets".

❼ From the producers' point of view, this means that it is no longer a smart strategy to leave oil in the ground on the assumption that a barrel pumped tomorrow will be worth more than a barrel pumped today. This might be one reason why not only Saudi Arabia, but also Russia is now producing at full capacity.

❽ Fourthly, Saudi Arabia is no longer the swing producer in the global oil market. It keeps producing at full tilt even though the oil price has collapsed. In the past, Saudi Arabia's balancing role meant that both low-cost and high-cost producers would supply the market at an elevated price

that guaranteed an income stream for all producers. No longer.

❾ Fifthly, with lower incomes, governments in oil producing countries can no longer lavish subsidies and other perks on their local populations. This might lead to domestic instability and repression—although it might also drive reforms in some producing countries desperately to shore up strained budgets and diversify away from oil.

❿ The countries that appear most immediately vulnerable to internal instability are Venezuela, Ecuador, Nigeria, Brazil, and the Central Asian producers like Azerbaijan.

⓫ Some people also worry about Saudi Arabia, which relies on oil for 70%–80% of its budget revenue. But the country still has over $600 billion in cash reserves and among the lowest production costs in the world. Like Russia, Saudi Arabia has tightened its grip on domestic politics as the oil price has fallen and its foreign policy has become more assertive. It has also shown signs of accelerating reforms at home.

⓬ Sixthly, technology is becoming more important. The boom in the U.S. shale industry has been driven by innovations in drilling technologies. If, as many claim, the U.S. is the new swing producer, the key role of technology makes it less predictable than Saudi Arabia, where decisions on production levels were centralized and political. Most energy analysts were surprised, for example, by how much the U.S. shale industry has been able to cut costs rather than production.

⓭ If the U.S. industry manages to maintain production at lower prices, America will be heading towards self-sufficiency in oil and gas. Its interest in guaranteeing stability in the Middle East might wane accordingly. This, in turn, might exacerbate the geopolitical rivalries between Saudi Arabia and Iran and other players in the region.

⓮ Seventhly, at the moment, the oil market is so nervous that geopolitical tensions lead to lower prices. Usually it is the opposite. This dynamic (together with the QE-induced search for yield) means that markets may not be pricing political risk correctly, and many businesses are not paying enough attention to it.

⓯ Lastly, since the big producers are pumping so much oil, they have less spare capacity. Saudi Arabia's spare capacity, for instance, has roughly halved since 2009. This could mean that any disruption—think of an escalation of Saudi-Iranian tensions, for example—could lead to a (temporary) price shock.

Activity Four Approach the Geopolitics of Oil and Gas

The following article will outline the 40 years' geopolitics of oil and gas so that you can have an overall view of how petroleum and geopolitics are intertwined. Read it carefully, classify the geopolitical events that lead to either the soaring or plummeting oil prices, and then share with your team members.

Reading C

Forty Years of Oil and Gas Geopolitics[1]

❶ The oil and gas markets are intrinsically linked to political and military developments worldwide, with trends influenced by such events as the wars in the Middle East, the Asian and subprime crises, and political upheaval in oil and gas producing regions. The major milestones from 1973 to today are described below.

October 1973—The Yom Kippur War

❷ The Yom Kippur War between Israel and a number of Arab states gave rise to what oil-importing nations termed the "first oil crisis". Here the Organization of Petroleum Exporting Countries (OPEC) used the "oil weapon" by proclaiming an embargo on countries supporting Israel. As a result, the price of crude rose fourfold from less than $15 a barrel to more than $50 (at 2011 prices).

1974–1980—The Counter-Attack by Oil-Importing Countries

❸ Following cost-cutting measures and rationing—a shock to the West after so many years of growth, oil-consuming countries reacted by exploring other forms of energy, like nuclear power. They also began developing new technologies that otherwise would have taken longer to get off the ground, paving the way for a production revival in mature U.S. oil fields and offshore production in the North Sea, for example. The result was a global rise in non-OPEC production. In 1974, the International Energy Agency (IEA) was founded with the initial goal of defending the interests of the oil-importing countries of the Organization for Economic Cooperation and Development (OECD).

1980–1981—The Iranian Revolution

❹ Oil prices doubled from mid-1978 to 1981 during the second oil crisis, as concerns over the Iranian Revolution and the Iran-Iraq War, combined with renewed growth in global consumption, pushed the price of oil up from $50 to $100 a barrel. This time, however, demand fell and technology continued to advance, setting the stage for a price decline in the following years.

1986—OPEC's Annus Horribilis

❺ With production continuing to rise in non-OPEC countries, the situation had become unbearable for the cartel by 1985. Saudi Arabia first attempted to stabilize plummeting prices by decreasing production. Then in December 1985, it performed an about-face by announcing that it would sell its crude oil unconditionally at market price in an attempt to claw back market share. The subsequent "oil glut" saw the price of crude collapse to as little as around $30. In late 1986,

1. Planete Energies. (2015, September 16). *Forty years of oil and gas geopolitics*. Retrieved from https://www.planete-energies.com/en/medias/close/forty-years-oil-and-gas-geopolitics

Saudi Arabia abandoned its failed offensive and allowed OPEC to re-establish production quotas.

September 1980–August 1988—The Long Iran-Iraq War

❻ Iran and Iraq, two of OPEC founding members, embarked on an interminable war that would cost nearly one million lives and inflict severe damage on both countries' oil facilities. Among the many root causes was Iraq's desire to increase its access to the Gulf and consequently boost oil exports.

August 1990–1991—The Iraqi Invasion of Kuwait

❼ Gasoline prices were also one of the main causes of the First Gulf War. Coming out of its bloody combat with Iran, Iraq invaded Kuwait after the emirate directly threatened to put further downward pressure on the price of crude. An international coalition forced the Iraqi troops out. The regime held on to power but had to withstand severe sanctions, including the "Oil-for-Food" Program that would remain in place until the Second Gulf War in 2003.

1998—The Asian Financial Crisis

❽ The Asian financial crisis took the world by surprise, and OPEC was no exception. After increasing production in 1997, the cartel took a severe blow when the crisis pushed oil prices down to historical lows of less than $20 a barrel. It reacted by slashing production quotas—with disastrous consequences for many oil-producing countries—until prices started to recover.

Late 1990s—The Rise of Liquefied Natural Gas (LNG)

❾ Towards the end of the century, liquefied natural gas (LNG) started to occupy a growing place in energy markets. Its share increased by around 7% a year from 2000 to 2012, and now accounts for some 10% of all gas consumed worldwide. And as LNG can be shipped and does not rely on land-based pipes for transportation, it has created a spot market for gas alongside long-term contracts between countries.

March–April 2003—The Second Gulf War

❿ The United States intervened in Afghanistan shortly after the terrorist attacks of September 11, 2001. Then, after accusing the regime in Bagdad of supporting al-Qaeda terrorists and harboring weapons of mass destruction, the U.S. invaded Iraq backed by an international coalition, albeit a smaller one than in 1990. The offensive and subsequent occupation destroyed the country's infrastructure (bridges, water treatment facilities, power plants, etc.), and brought Iraqi hydrocarbon production almost entirely to a halt. While OPEC intervention stopped oil prices from soaring during the initial intervention, a significant increase was observed in 2004.

2007 Onwards—The Shale Gas Revolution

⓫ After a long period of high conventional oil and gas prices, driven by a spike in global energy consumption, the United States began focusing its efforts on large-scale production of shale oil and gas. This marked the beginning of the "shale gas revolution".

2007–2008—Upward Price Spiral

⑫ Strikes in Venezuela, production stoppages in Nigeria, and other regional disruptions dragged global oil supply down. At the same time, tax cuts in the U.S. and increasing affluence in China and other emerging economies drove demand—and prices—to dizzying new heights. After selling for $96 in January 2008, the price of Brent crude soared to $144 in July of the same year.

Summer 2008—The U.S. Subprime Crisis

⑬ The collapse of the U.S. property market crippled the country's banking system before spreading to the rest of the world, sparking the worst economic crisis since the Great Depression. Global demand for oil contracted for the first time since 1982, falling 0.3% in 2008, just after OPEC had increased production. As a result, oil prices decreased to $35 a barrel.

2009–2014—The Arab Spring

⑭ Oil prices were volatile over the period as Iran threatened to block the Strait of Hormuz, a strategic oil shipping route, and the Arab Spring led to uprisings in several oil-producing countries. This was exacerbated by strong growth, and hence higher demand, in emerging economies. Consequently, oil prices rose steadily and squeezed economic growth worldwide. Some experts even warned of a "creeping oil crisis".

Mid-2014—Early 2015

⑮ Saudi Arabia grew anxious after the unconventional oil and gas boom in the United States began eating away at its market share. In an attempt to reduce the profitability of U.S. producers, it decided to maintain production levels in order to bring down oil prices. At the same time, slowing growth in China and crises in other countries such as Brazil caused demand to rise more slowly. Under the combined weight of these factors, the price per barrel fell to around $60 in early 2015, after having hovered well above $100 since 2011.

2015—Early 2016

⑯ The crisis is being dragged out by the standoff between Saudi Arabia and U.S. shale oil producers, which continue to maintain a high level of production. Even if oil demand in 2015 rose by 1.6 million barrels to 94.2 million barrels a day, as forecast by the International Energy Agency (IEA), there is still a large supply glut and the price of oil continues to decline. In January 2016, the price of Brent crude fell below $30 a barrel.

After-Class Activities

Activity One Conduct a Mini-Research Cooperatively

First read the following chart and figure out how the real oil price in 2015–2016 coincides

with what is analyzed in Reading C. Search for literature on the geopolitical analysis to account for this oil price plunge, hold a discussion with your team members about your findings, and then present them to the whole class. Also, you are asked to write down all the references you've searched for at the end of your report.

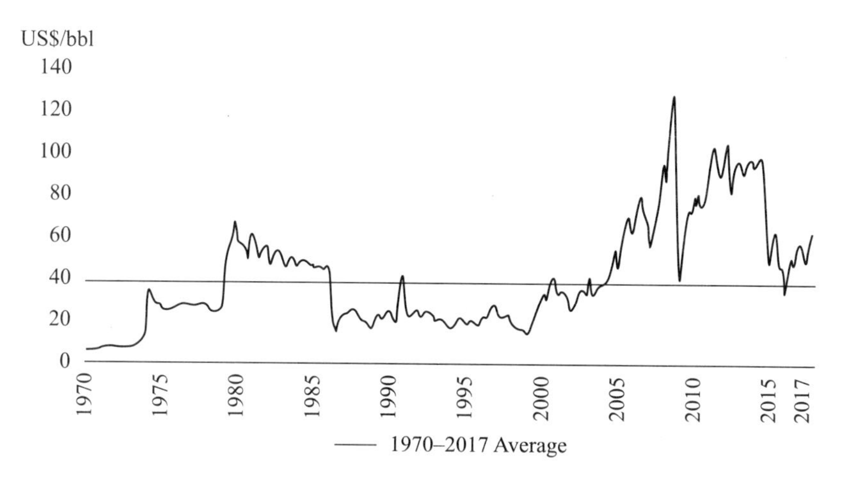

Figure 3-2 Biggest drop in oil prices in modern history

(Source: World Bank crude oil average. Last observation is November 2017.)

Activity Two Practice Summary Writing

Identifying important arguments is a shortcut to locate the writer's key points or opinions. The following is an in-depth analysis of effects of the oil industry on global politics. Read it carefully and then summarize the key points with not more than 150 words.

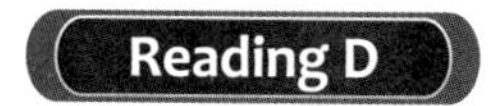

Reading D

The Oil Industry and Its Effect on Global Politics[1]

❶ Over the past century, modern society has developed a near unquenchable thirst for oil and after 100 years of searching and experimenting, there is still no reliable replacement.

❷ "Oil is Power!" I don't just mean power as in "energy"; I mean power, as in being a primary factor in the process of asserting and maintaining political dominance and control. Oil is needed

1. Oil Price.com. (2009). *The oil industry and its effect on global politics*. Retrieved from https://oilprice.com/Energy/Oil-Prices/The-Oil-Industry-And-Its-Effect-On-Global-Politics.html

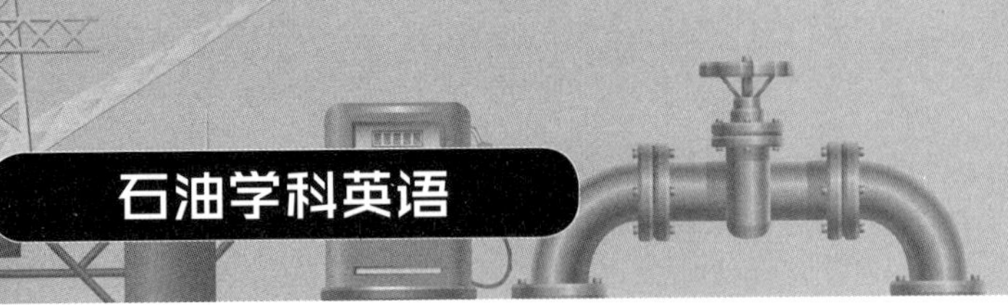

to grow food, build infrastructure, advance technology, manufacture goods and transport them to market. It lubricates the mechanisms of both national and international politics. Those who can consistently get their hands on the most oil, at the best prices...will rule!

❸ So what makes oil so highly valuable that individuals, companies, and sovereign states would actually be willing to go to war, if necessary, in order to defend or fight to win their "beloved"?

❹ First, "Oil is Universal!" It is a staple of our very existence! Oil plays a major role in practically every aspect of our lives from technology and transportation to the very food and business necessary for our survival.

❺ Second, "Oil is Unique!" While there may be various alternative energy supplies available for some industrial tasks such as creating electricity, there is currently no reasonable substitute for oil when it comes to transportation.

❻ Third, "Oil is Rare!" According to scientific calculations, oil is a progressively depleting fuel that is disappearing at an exponentially alarming rate. While there are still an undetermined number of rich, untapped oil deposits left to be discovered around the globe, reasonable arguments will continue as to just how quickly the world's oil supply might run out.

❼ However, even amongst the most optimistic and pessimistic prognosticators, there is virtually no debate that there is currently less oil available to us than there was just 50 years ago.

❽ As recently as the year 1900, coal accounted for 55% of the entire world's energy use while oil and natural gas contributed a mere 3% of the world's energy. One century later, coal provided only 25% of the planet's energy, natural gas has risen to 23%, and oil reigns supreme at just under 40%.

❾ Back in the year 2000, demand for oil was approximately 75 million barrels per day! Less than ten years later, the International Energy Agency now calculates that our global thirst for crude oil will actually double by the year 2030.

❿ Planes, trains, and automobiles, they all rely on oil. Whether it's driving the kids to school, hauling necessary foodstuff and commodities to market, or powering a warship, tank, missile launcher, or jet fighter in and out of battle zones, those who have oil prosper and those who don't...collapse!

⓫ So there is no surprise just how much international, geopolitical concern and conflict arise regarding oil and the companies that supply it around the globe. Over the years we've witnessed numerous rows being raised on the international scene, some merely escalating into confrontations quelled by "quid pro quo" agreements while others having led to boycotts, United Nations, censures and in some cases invasions and all out wars!

⓬ Throughout history, there have been numerous, seriously contentious conflicts about oil

involving the United States, Russia, the former Soviet Union (particularly Ukraine), Turkey, Britain, Germany, Norway, the Netherlands, France, Italy, Japan, Saudi Arabia, Iran, United Arab Emirates, Afghanistan, Kuwait, Iraq, Mexico, Venezuela, Indonesia, Nigeria, Algeria, and Libya, just to name a very few of the many sovereign principalities and geographical locations that have found reason to "come to odds" and on occasion, "to arms", over oil.

⑬ Many of the most prosperous countries also tend to be those countries who have made "arrangements" to consistently receive large supplies of life-giving oil, at reasonably low oil prices, for an extended period of time. These entities that "have" quite naturally don't want to go without and will often be willing to use whatever political might they find necessary to protect their position of prominence.

⑭ On the other side of the coin, higher oil prices have also served to bring greater political stability and prosperity to several regions around the planet. Some of these locations, including Mexico, Columbia, Venezuela, China, India, several of the Persian Gulf States, Russia, as well as many former Soviet Central Asian Republics and portions of the continent of Africa, particularly Nigeria, are just getting their first tastes of "the good life" and are quickly developing a strong liking to the flavor.

⑮ For some countries, higher oil prices mean finally having the money needed to invest in desperately outdated infrastructure, technology, and means to successfully building a sustainable defense and military that protects the borders and sovereignty of the nation, eliminating many incursions, invasions and all out turf wars before they can ever get started. People who feel safe tend to prefer the sweet fruit of peace!

⑯ The old axiom has never been more true: "As flows the oil, so flows prosperity." Everything from a country's economy and currency exchange rate to their population's overall sense of security and political stability seems to hinge precariously on what has come to be known as "black liquid gold"!

⑰ The very political success or failure of any ruling regime and the very survival of its citizens is dramatically affected, not simply by the mere possession of oil, but by effectively controlling the price of this all important fuel.

⑱ One thing that nearly all governments seem to agree upon is the importance of maintaining stability in both the market and ability of oil to reach those energy thirsty nations that it serves.

⑲ Meanwhile, there are strong proponents of various political agendas hoping to alter the landscape of various regions, whether they are agents of Democracy, throwbacks to the days of socialism and communism, or an ever expanding "universal industrialism" that crosses all borders and nationalities.

⑳ No one can possibly know for sure what the future holds, but one thing is absolutely for certain. For the next 50 to 100 years, oil will continue to play a major role in determining the geopolitical make-up of this planet. Whether the international game being played is based on economics as in "monopoly" or world domination by way of military prowess, such as in "risk", the one common factor will be the oil that lubricates the wheels of progress towards prosperity and political power!

Integrated Exercises

I Read the following academic words, and check whether you can use them appropriately. For those you can not, look up in a dictionary or search online about their contextual use. Write down notes to strengthen your memory.

Reading A

1. indispensable
2. evident
3. conflict
4. sway
5. sentiment
6. safeguard
7. tactic
8. backfire
9. initiative
10. given (*prep.*)
11. revenue
12. accomplish
13. demonstrate
14. erect
15. exploitation
16. lessen
17. escalate
18. landmark
19. thwart
20. thrust
21. ally
22. surrender
23. uninterrupted
24. ultimate
25. essence
26. historic
27. coincidence

Reading B

1. undermine
2. revive
3. coordinate
4. uncertainty
5. assumption
6. tilt

7. collapse
8. subsidy
9. perk
10. repression
11. diversify
12. vulnerable
13. internal
14. grip
15. assertive
16. boom
17. innovation
18. elevated
19. predictable
20. centralized
21. analyst
22. self-sufficiency
23. guarantee
24. wane
25. accordingly
26. rivalry
27. dynamic
28. roughly
29. halved
30. disruption
31. escalation

Reading C

1. intrinsically
2. milestone
3. rationing
4. initial
5. renew
6. unbearable
7. unconditionally
8. subsequent
9. offensive
10. embark
11. interminable
12. inflict
13. boost
14. troop
15. withstand
16. slash
17. intervene
18. harbor
19. soar
20. intervention
21. spike
22. stoppage
23. affluence
24. dizzy
25. volatile
26. uprising
27. squeeze
28. creeping
29. profitability
30. hover
31. standoff

Reading D

1. assert
2. maintain
3. manufacture
4. lubricate
5. mechanism
6. consistently
7. survival
8. substitute
9. undetermined
10. currently
11. mere
12. supreme
13. approximately
14. automobile
15. hauling
16. foodstuff
17. warship
18. prosper
19. concern
20. witness
21. confrontation
22. quell
23. boycott
24. censure
25. contentious
26. involve
27. prosperous
28. prominence
29. portion
30. flavor
31. outdated
32. sovereignty
33. incursion
34. axiom
35. hinge
36. possession
37. proponent
38. agenda
39. landscape
40. agent
41. throwback
42. monopoly
43. domination

II Decide on the contextual meaning of the following terms and expressions.

Reading A

1. political weapon: ______
2. price at the pump: ______
3. safeguard oil reserves: ______
4. enact strict measures: ______
5. public relations disaster: ______
6. the inclement weather: ______
7. industrialized era: ______

8. open the chapter of oil-politics:________________

9. in return for preferential treatment:________________

10. untapped oil fields:________________

11. recognize the strategic importance of oil:________________

12. atomic bomb:________________

13. unconditional support:________________

Reading B

1. raise a host of political questions:________________

2. oil glut:________________

3. scarce resource:________________

4. geopolitical battle:________________

5. global market share:________________

6. lift sanctions:________________

7. supply and demand:________________

8. reach a climate change target:________________

9. stranded asset:________________

10. produce at full capacity:________________

11. produce at full tilt:________________

12. the swing producer:________________

13. an income stream:________________

14. shore up strained budgets:________________

15. budget revenue:________________

16. cash reserves:________________

17. internal instability:________________

18. tighten grip on domestic politics:________________

19. drilling technology:________________

20. self-sufficiency in oil and gas:________________

21. geopolitical tensions:______

22. spare capacity:______

23. price shock:______

Reading C

1. subprime crisis:______

2. political upheaval:______

3. oil-importing nations:______

4. oil-consuming countries:______

5. pave the way for a production revival:______

6. non-OPEC production:______

7. plummeting price:______

8. production quota:______

9. root causes:______

10. push oil prices down to historical lows:______

11. the "Oil-for-Food" Program:______

12. a spot market:______

13. terrorist attack:______

14. weapon of mass destruction:______

15. water treatment facilities:______

16. an international coalition:______

17. shale gas revolution:______

18. emerging economies:______

19. property market:______

20. a strategic oil shipping route:______

21. the unconventional oil and gas boom:______

22. supply glut:______

23. the price of Brent Crude:______

Reading D

1. political dominance and control:________________
2. sovereign state:________________
3. the mechanisms of international politics:________________
4. untapped oil deposit:________________
5. reign supreme:________________
6. missile launcher:________________
7. jet fighter:________________
8. battle zones:________________
9. build a sustainable defense and military:________________
10. geopolitical concern and conflict:________________
11. escalate into confrontations:________________
12. geographical locations:________________
13. political might:________________
14. position of prominence:________________
15. political stability and prosperity:________________
16. the borders and sovereignty of a nation:________________
17. turf wars:________________
18. the overall sense of security:________________
19. political agendas:________________
20. the geopolitical make-up:________________
21. military prowess:________________

II Analyze the grammatical structure of the following complex sentences, figure out the meaning of each sentence, and paraphrase them.

1. Oil and politics have always gone together for a simple reason; since oil became an

indispensable commodity without which the world as we know it today would not function, countries that produce oil have learned how to use it as a weapon. (Reading A, Para. 1)

2. Hoping to sway Western sentiments in favor of the Arab causes Arab oil producing countries such as Saudi Arabia and the Gulf sheikdoms agreed to reduce their output. (Reading A, Para. 2)

3. The tactic employed by the oil producers however backfired: Forced by some governments to leave their cars in the garage on alternate days along with having to pay more money for less gas, the 1973 Arab oil embargo initiative was a public relations disaster. (Reading A, Para. 3)

4. There were alternatives to Arab oil except that until the Arab embargo of 1973 purchasing Arab oil was far less expensive than erecting platforms in the inclement weather of the North Sea, for example, in Norwegian waters or off the English coast. (Reading A, Para. 5)

5. A more recent reason might be to prevent its regional rival Iran from re-entering the oil market, now that sanctions have been lifted. (Reading B, Para. 3)

6. From the producers' point of view, this means that it is no longer a smart strategy to leave oil in the ground on the assumption that a barrel pumped tomorrow will be worth more than a barrel pumped today. (Reading B, Para. 7)

7. This might lead to domestic instability and repression—although it might also drive reforms in some producing countries desperate to shore up strained budgets and diversify away from oil. (Reading B, Para. 9)

8. If, as many claim, the U.S. is the new swing producer, the key role of technology makes it less predictable than Saudi Arabia, where decisions on production levels were centralized and political. (Reading B, Para. 12)

9. The oil and gas markets are intrinsically linked to political and military developments worldwide, with trends influenced by such events as the wars in the Middle East, the Asian and subprime crises, and political upheaval in oil and gas producing regions. (Reading C, Para. 1)

Unit 4

Petroleum and Technology

Unit Objectives

Goal 1

Get informed of how technology is promoting the oil and gas industry through task-based activities.

Goal 2

Get to know how the current technological advances are reshaping the petroleum industry and meeting the global energy demand.

Goal 3

Get a glimpse of the future development of the oil and gas industry through the perspectives of the academia.

Petroleum science has evolved from undeveloped geology to supercomputer-based calculations and 3D views of the subsurface. It has taken the drilling process from a guessing game to the defined targeting of fields. The 21st century oil and gas industry is charged by innovation and technology. It has dramatically altered the manner in which oil and gas reserves are identified, developed, and produced. Advancements in technology have also improved environmental protection and conservation of natural resources.

Pre-Class Activities

Activity One Search for the Definitions

Search online for the definitions of the following terms or concepts and share your findings with your team members.

1. proven reserves:____________________
2. seismic data:____________________
3. remote sensing:____________________
4. horizontal drilling:____________________
5. hydraulic fracturing:____________________
6. research and development (R & D):____________________
7. unconventional oil and gas:____________________
8. upstream/midstream/downstream:____________________
9. crude oil refining or natural gas processing:____________________
10. digital oilfield:____________________
11. COO:____________________

Activity Two Watch the Video About Oil Drilling Innovation

Watch the following video to learn how horizontal drilling has changed the global energy landscape and summarize the key points of the impacts it has had on both ecology and economy.

 **Video:** Horizontal Drilling Is Changing the Game?

While-Class Activities

Activity One Make a Seminar Discussion

Hold a discussion with your team members about the possible areas where R & D in the

oilfield might involve. Take the horizontal drilling you've learned from the video above as an example to further illustrate how R & D can improve the oil production efficiency. Total R & D expenditures are divided as follows according to INTEK:

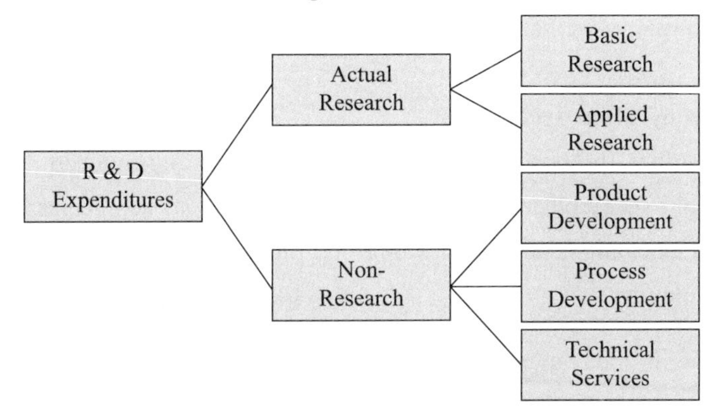

Activity Two Get to Know Technological Advances

Read the following article to learn how technology can affect the oil and gas industry. Summarize the key points and present them to the whole class.

Reading A

How Technology Is Saving the Oil and Gas Industry[1]

Andrew Medal

Technology is helping revive and transform legacy industries by creating efficiencies where needed.

❶ To say the oil and gas industry has had a tough couple of years would be a bit of an understatement. Today, public perception is that oil and gas companies are not keeping up with the times. Not only are organizations, like Google, spearheading clean energy projects that threaten oil and gas, but other industries are also adopting new technology at faster paces.

❷ From robotics to artificial intelligence, organizations across verticals recognize that in order to stay relevant among consumers, they have to change with the times. So why are oil and gas companies still lagging behind?

1. Medal, A. (2017, September 29). How technology is saving the oil and gas industry. *Inc*. Retrieved from https://www.inc.com/andrew-medal/how-technology-is-saving-oil-gas-industry.html

❸ Luckily, there's still time for companies within these fields to revamp their operations and propel themselves to the modern-day playing field. All organizations existing in traditional verticals can learn how to reignite progress within their companies by assessing the problems that plagued oil and gas, as well as the technological advancements that have the potential to turn things around.

The Problem: Slow to Adopt Technology

❹ Like many other industries, the oil and gas industry is undergoing a rapid digital transformation. Unlike many other industries, however, oil and gas companies have been slow to adopt technological innovations on the software front. Lloyd's Register accurately summarizes the current state of the industry:

"Innovate or die. In today's energy industry, the speed of adoption lags behind that of other industries; industries that are subject to the same rash of safety, legal, commercial, and financial pressures faced by Energy companies, such as Aerospace."

❺ Many of the reasons that the industry is slow to evolve are understandable; the capital that was once plentiful has seen a rapid decline and the industry is stuck in the "old way" of doing things. Discussing the current state of the industry, Shiva Rajagopalan, CEO of Seven Lakes Technologies, a mobile field data capture platform working in the oil and gas industry, says:

"Ironically, the reason oil companies have been slow to invest in more efficient practices is due to their wealth. High market prices and a steady influx of capital allowed them to hide inefficiencies. The focus was growth and reserves exploration. If they are killing it in the market, why should they change? There is an old-guard that doesn't necessarily understand how technologies can drive bigger business value."

❻ While this enormous undertaking may scare others, entrepreneurs should view this as a massive opportunity.

The Opportunity: Digital Transformation

❼ The oil and gas industry, combined, are multi-billion dollar industries; big oil is big money and the time is now to bring the industry into the digital age.

❽ The following are top technologies that have been adopted in other industries, yet have been slow to integrate into the oil and gas industry.

Robotics

❾ The industry was slow to adopt the use of robotics before 2014, but now companies climbing out of the collapse are implementing them. For example, Iron Roughneck, which was developed by a company called National Oilwell Varco Inc., automates the dangerous and

repetitive tasks of connecting drill pipes on oil rigs.

⑩ While technologies like these do add value to the companies, there is potential for workers to lose their jobs, which is a recurring fear that many have when it comes to robots.

⑪ Chris Blackford, the founder of Sky-Futures, a drone company for the oil and gas industry, explained that "the inspection data we can collect in five days take rope-access technicians about eight weeks."

Artificial Intelligence

⑫ In the oil and gas industry, AI allows companies to uncover trends that pinpoint and predict inefficiencies. Leveraging AI to improve performance operations from C-level to field workers, automate processes, streamline manual business operations, and connect with IoT devices, makes every arm of the company more efficient and profitable.

⑬ "To survive the current era of cheap oil, we will see the democratization of tools like AI, automation, and IoT. The oil and gas sector must capitalize on such business intelligence; otherwise, they will undoubtedly be left behind in a worldwide digital revolution," says Rajagopalan.

Cloud Computing

⑭ As the oil and gas undergo this enormous transformation to a digital infrastructure, cloud computing will prove to be a powerful engine. The sheer amount of data companies can harness and further analyze through automation, will reduce operational expenses, down well times, and lessen risk. As more oil and gas companies integrate cloud computing, this will empower field workers to optimize production.

⑮ "To take effective action, the entire production chain, from COO right down to on-site well engineers, needs to see the very detailed cost and production data, narrowed down to the invoice level. By leveraging cloud computing capabilities, accuracy and transparency are achieved in the shortest amount of time to drastically improve well-cost management," explains Rajagopalan.

⑯ Even industries that aren't traditionally progressive can no longer afford to staunchly opposed change and digital transformation. Luckily for organizations that may have been slow to adapt, emerging technologies, including cloud computing, AI, and robotics can be easily implemented and impactful almost immediately.

Activity Three Get to Know Key Technological Drivers

Following your summary in Reading A, run through the following article for what technology might hold for the oil and gas industry. You are required to share your comprehension of at least one of the discussed trends with your team members.

Reading B

Five Trends in Oil and Gas Technology[1]

Mac Elatab

❶ Whether we like it or not, hydrocarbon fuels are not going away anytime soon, and innovations in oil and gas tech have the potential to impact everyone. If technology makes oil and gas easier, safer, cleaner, and cheaper to extract, energy prices and quality of life could improve for everybody. And if that doesn't appeal to the better angels of your nature, it presents a huge business opportunity.

❷ Here's a look at the five key innovation areas the extraction industry is focused on—and where the biggest opportunities lie for oil and gas technologists:

Software "Eats" Oil and Gas: The Digital Oilfield and Beyond

❸ As Marc Andreessen has written, software is "eating the world", and oil and gas is no exception. "The digital oilfield" is something of a buzzword in the oil and gas industry—tossed around like "the cloud" or "big data" are in IT. The basic premise is a web-based visualization platform from which companies can manage, measure, and track all of the data coming from all over the oilfield.

❹ Finding Petroleum has described the digital oilfield as an oilfield where "all the components integrate and communicate as well as your body does". A McKinsey report describes a "digital oil-field" where "instruments constantly read data on wellhead conditions, pipelines, and mechanical systems. That information is analyzed by clusters of computers, which feed their results to real-time operations centers that adjust oil flows to optimize production and minimize downtimes."

❺ This represents a big opportunity for oil and gas companies. According to Oil and Gas Investor, total upstream energy IT support spending is about 25 cents per barrel of oil. Booz & Co. mentioned that some analysts believe digital oilfield technologies could increase the net present value of oil and gas assets by 25%.

❻ A digital oilfield can also minimize the impact of the lack of qualified skilled labor; Booz & Co. predicts that the labor gap could reach one million by 2015. The McKinsey report describes how "one major oil company" was able "to cut operating and staffing costs by 10%–25% while

1. Elatab, M. (2012, March 28). Five trends in oil & gas technology, and why you should care. *VentureBeat*. Retrieved from https://venturebeat.com/2012/03/28/5-trends-in-oil-gas-technology-and-why-you-should-care/

increasing production by 5%".

❼ Major software companies have jumped on this trend. EMC announced a development center in Rio de Janeiro to focus on big data analytics for oil and gas. In fact, EMC boasts that more than 95% of Forbes Global 2000 oil and gas companies use its technology. Other large IT and software companies with solutions designed for oil and gas companies are IBM, Microsoft, Progress Software, Cisco, Wipro, and SAS. Schlumberger and Baker-Hughes, two major oilfield services companies, have significant digital oilfield practices. A smaller company in this space is Chevron Technology Ventures—backed Mobilize, which enables energy firms to aggregate data in real time from multiple vendors.

❽ A handful of other venture-backed software companies are tackling oil and gas. Some small companies in the space include Transform Software and Services, a developer of visualization and interpretation software designed for geophysical, geological, and engineering workflows. Five Cubits provides tracking software and plant control systems for bulk material companies, including petroleum, coal, and minerals companies. Rock Flow Dynamics provides software tools for reservoir engineering, specifically a reservoir simulator.

Accessing the Previous Inaccessibility

❾ When Marco Polo visited Azerbaijan, he found oil gushing out of the ground. Today, oil is considerably harder to find, but new technologies are making the once-inaccessible accessible. For example, BP originally estimated it would only be able to recover 40% of the oil at Prudhoe Bay in Alaska but has since revised this upward to 60%. Some of the new techniques BP has used include fracturing the rocks (applying pressure to the rocks to create fine cracks that can stimulate the flow of trapped natural gas), injecting low salt water into a reservoir to push out oil trapped in rock pores, injecting carbon dioxide into wells, and feeding in certain micro-organisms to help oil flow.

❿ VC-backed companies, including Oxane and Glori Energy, have sought to facilitate these types of operations. Oxane has commercialized technology from Rice University to create new ceramic proppant to increase the rate of production and the total amount recovered, and to reduce the environmental impact of fracturing. (Proppants are particles added to fracturing fluid in order to hold fractures open.) Glori Energy improves the sustainability and efficiency of recovering oil trapped in reservoirs by stimulating naturally occurring microbes. For more challenging jobs, Foro Energy has developed long-distance laser technology to destroy ultra-hard (high compressive strength) rocks.

⓫ When turning from oil to coal, some coal is too deep to mine. Laurus Energy, a developer of underground coal gasification projects, is able to convert this goal into gas streams, which can be used as feedstocks or to produce low energy carbon.

Working in Remote Environments

⑫ Oil and gas exploration today is based in some of the most remote places in the world. It is hard to imagine places more inaccessible than Prudhoe Bay (on the North coast of Alaska), Russkoe Field (above the Arctic Circle in Russia), the Campos Basin (offshore, Brazil), or the Tengiz in Kazakhstan. North Sea oil is the reason Aberdeen has the busiest heliport in the world. Oil and gas companies have looked to technology solutions to manage these distances. For example, RigNet is a venture-backed company focused on IT networks for drilling rigs and offshore vessels. NuPhysica, a telemedicine company, solves a major need within the oil and gas industry of handling employee injuries when they're out in the middle-of-nowhere. The company uses advanced videoconferencing solutions to allow doctors to diagnose and potentially treat patients remotely. In one case, it saved an oil company $30,000 by enabling an injured worker to be diagnosed on an oil rig rather than emergency transported to shore. Even mundane things, such as remote collaboration software, are crucial. VSEE Labs, a Salesforce.com-backed startup, provides video conference and collaboration software used by Shell Oil and Saudi Aramco.

Minimizing the Harm of Hydrocarbons

⑬ Cleantech investing may not be dead, but alternative energy is arguably in the "trough of disillusionment" in the hype cycle following recent scandals and the disappointing financial performance of public cleantech companies. The WilderHill Clean Energy Index is down 55% since it was launched in 2004. This has increased interest in making production and consumption of hydrocarbon fuel less harmful, rather than making alternative fuels commercial. For example, companies such as PWAbsorbents and GeoPure HydroTechnologies have tackled the problem of treating water produced in the extraction of natural resources.

⑭ Other companies focus on decreasing demand rather than trying to increase supply. Cerion Energy produces a nanotechnology-based diesel fuel that decreases consumption by a minimum of 8%, while reducing emissions.

Turning Lemons into Lemonade and Fuel Metamorphoses

⑮ There is another suite of companies that try to take one resource and turn it into another, more valuable or convenient resources. Fractal Systems is a VC-backed technology that upgrades heavy oil and bitumen. Heavy oil and bitumen have greater densities than light crude oil; they are more expensive to refine and produce more pollution; however, there is much more heavy oil and bitumen in the world than light crude.

⑯ Siluria Technologies is a company that replaces oil with natural gas for manufacturing processes. The advantage of doing this is that natural gas is cheap and abundant in the U.S. (and has become cheaper and more abundant because of fracking technology) and pollutes less than oil (0.23 $kgCO_2/kWh$ for natural gas vs 0.27 $kgCO_2/kWh$ for gasoline). In a similar vein, Ciris Energy

converts coal to natural gas. This is valuable, because coal is the most abundant fossil fuel in the U.S. and China, but converting it to natural gas makes it more environmentally friendly (0.23 $kgCO_2$/kWh for natural gas vs 0.37 $kgCO_2$/kWh for coal).

⑰ Even more audacious are companies turning waste into energy. Enerkem is pioneering technology to use garbage to replace petroleum. SunCoal Industries turns organic waste, such as garden compost, straw, and chicken manure, into carbon-neutral coal. Agilyx developed technology to convert plastics into synthetic crude and other petrochemicals. Plastic is heated until it turns into a gas, which is condensed into a liquid, and then the hydrocarbons are separated.

After-Class Activities

Activity One Watch the Video About the Emerging Trends

Watch the following video and take a summary note on the trends discussed.

 **Video:** Headlines Special Report: 2017 Emerging Oil, Gas Trends

Activity Two Practice Summary Writing

Subtitles can aid you a great deal in writing a summary. The following is an excerpt of an article, presenting expert opinions on what is likely to shape the future of oil and gas. Read it carefully and then summarize the key points with not more than 200 words.

The Future of Oil and Gas: Eight Bold Industry Predictions[1]

Talal Husseini

❶ As oil prices tentatively recover from the 2014 crash and investments in alternative renewable energy sources gain momentum, oil and gas companies need to innovate to stay

1. Husseini, T. (2018, August 03). The future of oil and gas: Eight bold industry predictions. *Offshore Technology*. Retrieved from https://www.offshore-technology.com/digital-disruption/blockchain/the-future-of-oil-and-gas-predictions/

competitive and keep the fuel flowing. *Offshore Technology* asks industry experts for their insight into how technological advancements will shape the future of oil and gas. Industry experts give their opinions on what is likely to shape the future of oil and gas.

The Future of Oil and Gas: "Smart Drilling"

❷ Nowadays, it seems like more and more companies want to become the Carl Lewis or Usain Bolt of drilling. Get out the blocks fast, hit every stride sweetly, and cross the finish line to first oil in record time.

❸ As any elite runner will tell you, the equipment alone doesn't win you the race. Equally, if not more important, is developing a race plan, road-testing that plan, and developing the intelligence to know exactly when, where, and how to hit the gas.

❹ So when it comes to the future of the oil and gas industry, "smart drilling" will be the key and require a combination of technology and thinking that reimagines how firms manage and execute a more harmonized approach to early well life.

❺ The key is ensuring that design, analysis, equipment selection, and implementation are all aligned and buttressed by operational expertise. Where companies lack the expertise or resource, initiation specialists will fill the void.

❻ As drilling projects grow in ambition, smarter equals faster. By combining integration and intelligence through specialist providers in the initiation phase with best-in-class technology, "smart drilling" promises to give projects the solid footing needed to keep the industry running for decades to come.

—James Larnder, Managing Director, Aquaterra Energy

The Future of Oil and Gas: Incorporating Blockchain

❼ One technology set to transform the oil and gas sector is blockchain. In fact, the blockchain revolution is starting here and now. The real task for the oil and gas sector is how quickly it can move to take advantage of the many opportunities that blockchain will bring.

❽ For oil and gas businesses, data have gone from an asset to a burden. Companies are drowning in data and urgently need a way to control and authenticate information. Blockchain has enormous potential to reduce the risk of fraud, error, and invalid transactions in energy trading, make financial transactions more efficient, facilitate regulatory reporting requirements, and enable interoperability.

❾ Blockchain will have huge benefits both upstream and downstream. From scheduling equipment maintenance to managing exploration acreage records, blockchain offers a single, unalterable record of transactions and documentation between numerous parties. Distributed ledgers also create more efficient and transparent downstream activities, such as exchanging products, secondary distribution delivery documentation, demurrage, and claims management. As

for midstream, it will revolutionize joint ventures, risk management, contracting, and regulatory compliance.

⑩ The possibilities of blockchain in oil and gas have few limits—and we're yet to see more than a glimpse of its full capabilities.

—Simon Tucker, Head of Energy and Commodities, Infosys Consulting

⑪ The energy sector is seen as the next frontier for blockchain development outside the financial sector, where the distributed ledger technology has had its biggest impact to date. Blockchain is critical to unlocking the efficiency potential of distributed energy generation and disintermediating the public and private utility companies. So does blockchain open up efficient fundraising through initial coin offerings (ICOs).

⑫ More than 1,500 ICOs have taken place in the energy space over the last two or three years. Admittedly, a disproportionate number of these token offerings have been electricity or renewables-focused, but the number of token offerings in the traditionally technologically phobic oil and gas sector is now rising.

⑬ We have already seen strong interest in our own ICO for an onshore hydrocarbon concession and another standout example of an ICO in the sector is WePower, a Lithuanian-based green energy trading platform, which raised €32m ($40m) in February 2018—the largest ICO in the energy sector to date.

⑭ Distributed ledger technology could also see the advent of peer-to-peer energy trading, as demonstrated by Power Ledger, which allows consumers to buy and sell clean solar energy, disrupting the established norms of energy provision.

—Robert Pyke, CEO, Aziza

The Future of Oil and Gas: Blurring the Lines Between Fossil and Renewable Fuels

⑮ Liquid fuels are still difficult to replace and while their reliance will be reduced as they get supplemented by biofuel and electrical energy sources, it will be a number of decades before they are phased out completely.

⑯ The prices of oil and gas will be perpetually lower for the foreseeable future as fracking will gradually open up more sources of cheap production, while demand slowly falls with the adoption of more renewables.

⑰ Better renewable energy technology and sources will eventually replace the use of oil as a combustible fuel but this will free it and other sources such as coal to be used to produce more sophisticated carbon products, hydrocarbons, and polymers, making it a feedstock rather than a fuel.

⑱ Finally, the ability to chemically synthesize oil and gas from more types of natural materials will blur the line between renewable and fossil fuels to the extent where it becomes a forgotten issue. The ability to synthesize these products will mean that even fossil fuels can be readily

replaced so the market will drive the source again.

—Michael Martella, CEO, Anergy

The Future of Oil and Gas: A "Gig Economy"?

⑲ Despite the market's challenging period, there won't be a Kodak moment. Oil and gas isn't going anywhere and the reality is that the transition to 100% renewable energy use in the U.K. won't happen in our lifetimes. Globally, emerging countries will also want to capitalize on their oil reserves—providing, literally and figuratively, a pipeline of growth for the future.

⑳ The industry will also become more collaborative. The billions of pounds spent on exploration and building platforms in new oilfields will be shared amongst multiple industry backers. As a whole, the oil and gas industry will become significantly more risk-averse, with companies working on joint ventures in order to avoid another big downturn.

㉑ The aversion to risk will filter into the organizational culture, with companies looking at how they can run leaner and meaner operations. Having project teams sitting around waiting in the wings for a new assignment will no longer exist. The industry will rely more and more on flexible workers to be brought in for specific projects. Expect the "gig economy" to come to the oil and gas industry.

㉒ Technology will be a facilitator in the transformation of organizations. The future of oil and gas is unmanned platforms, with workers transitioning from offshore to onshore office-based roles. Generalist manager roles will die out as the demand for short-term, niche skill sets to implement IT systems and bring oilfields "online" grow.

—Terry Noble, Lead Consultant for the Energy and Utilities Practice, Odgers Interim

The Future of Oil and Gas: "Smart Oilfield" Technology

㉓ All oil and gas operators, wherever they are located, would like to focus more of their budgets and efforts on improving productivity whilst at the same time monitoring their local and remote assets dynamically. This is only possible via the deployment of a truly "smart oilfield" technology, which is able to provide all critical data in real-time without any downtime.

㉔ From a technology standpoint, the ideal solution would need to seamlessly connect all systems and hardware platforms across the various fields of operation, integrating exploration, drilling and production facilities, and ultimately delivering useful data and video streams to a central location, allowing the operators to make better and quicker decisions.

㉕ One of the challenges faced by the oil and gas industry is related to the fact that contractors and assets generally move from one location to another on a daily or weekly basis. Additional challenges are directly linked to environmental conditions such as extreme temperatures and frequent sandstorms. Providing uninterrupted service without the involvement of on-site technicians after the move of a rig or drilling platform, for example, is only achievable through the adoption of high capacity wireless platforms, such as InfiNet's, which are able to auto-align and

mitigate these challenges.

㉖ The main factors that will drive the further adoption of wireless technologies in this industry sector will be the increasing demand for smarter sensors/devices in the field, as well as the desire for managers to stay continuously connected with their valuable assets.

—Kamal Mokrani, Global Vice President, InfiNet Wireless

The Future of Oil and Gas: The Challenge of Rig Decommissioning

㉗ One of the biggest challenges facing oil and gas companies is the cost of decommissioning aging rigs around the world, a toll which will reach $13 billion a year by 2040, with some set to be even more expensive than that. More than 600 rigs need to be decommissioned by 2021 and the most straightforward option—simply sinking the rigs—is not feasible. Oil companies need to come up with environmentally friendly ways of decommissioning rigs or face a potentially huge backlash from increasingly environmentally-conscious consumers.

㉘ The problem is that predicting and addressing the environmental impacts of various decommissioning methods is complex. There are always competing interests and trade-offs to be considered with circumstances varying from project to project. For example, the most carbon-neutral option in one instance might be unacceptable from a health and safety perspective in another.

㉙ Tackling the problem demands that geoscientists and engineers make confident, data-driven decisions, using the most relevant and accurate research available. Therefore, in order to manage the huge challenge of decommissioning, companies have to ensure they are providing their engineers with accurate and trusted information platforms, allowing them to be as efficient as possible in their work and make confident decisions.

—Phoebe McMellon, Director of Oil and Gas Strategy, Elsevier's R & D Solutions

The Future of Oil and Gas: Digital Transformation Offshore

㉚ The oil and gas industry is "always-on" and has long been defined by the legacy systems that help it to function. However, as digitalization continues to transform this sector, organizations are looking for common technologies that can help them balance requirements for uptime, security, and safety with the need to take advantage of digital innovation.

㉛ Digital transformation does not require a "rip and replace" approach. Instead, organizations should view this as an opportunity to improve the functional capabilities of their facility and move to a new software environment, which extends the life of the traditional legacy systems.

㉜ No one can deny that digital technologies are the future. Looking ahead, we can expect to see many companies turn to open standards to help them improve operational efficiency and grapple with digital complexity. Organizations within the oil and gas sector are working harmoniously with their peers to create open systems that will ensure digital transformation initiatives can be done at a low cost and with very little disruption. When there is a safe path to digital for these companies, it will unlock significant cost savings and efficiency for the wider process automation industry.

—Ed Harrington, Director, the Open Group Open Process Automation Forum

Integrated Exercises

I Read the following academic words, and check whether you can use them appropriately. For those you can not, look up in a dictionary or search online about their contextual use. Write down notes to strengthen your memory.

Reading A

1. understatement
2. propel
3. reignite
4. plague
5. influx
6. inefficiency
7. entrepreneur
8. automate
9. repetitive
10. recurring
11. pinpoint
12. leverage
13. streamline
14. harness
15. sheer
16. invoice
17. transparency

Reading B

1. buzzword
2. premise
3. instrument
4. optimize
5. qualified
6. analytics
7. boast
8. aggregate
9. venture-backed
10. visualization
11. geophysical
12. geological
13. workflow
14. bulk
15. inaccessible
16. gush
17. originally
18. inject
19. facilitate
20. commercialize

21. particle
22. sustainability
23. ultra-hard
24. compressive
25. gasification
26. feedstock
27. remote
28. telemedicine
29. diagnose
30. mundane
31. collaboration
32. crucial
33. startup
34. arguably
35. index
36. disillusionment
37. scandal
38. launch
39. tackle
40. minimum
41. density
42. suite
43. upgrade
44. vein
45. audacious
46. pioneer (*vt.*)
47. compost
48. straw
49. manure
50. synthetic
51. condense

Reading C

1. bold
2. tentatively
3. momentum
4. elite
5. execute
6. initiation
7. harmonize
8. implementation
9. align
10. buttress
11. ambition
12. void
13. authenticate
14. invalid
15. interoperability
16. acreage
17. phase
18. documentation
19. fraud
20. compliance
21. frontier
22. offering
23. fundraising
24. admittedly
25. disproportionate
26. phobic
27. concession
28. standout
29. advent
30. peer-to-peer
31. disrupt
32. reliance

33. perpetually
34. foreseeable
35. combustible
36. sophisticated
37. synthesize
38. blur
39. transition
40. emerging
41. capitalize
42. figuratively
43. collaborative
44. amongst
45. backer
46. risk-averse
47. downturn
48. aversion
49. facilitator
50. unmanned
51. assets
52. downtime
53. standpoint
54. seamlessly
55. adoption
56. ultimately
57. mitigate
58. toll
59. straightforward
60. feasible
61. backlash
62. environmentally-conscious
63. address (*vt.*)
64. trade-off
65. carbon-neutral
66. legacy
67. digitalization
68. grapple

II Decide on the contextual meaning of the following terms and expressions.

Reading A

1. organizations across verticals:____________________
2. the modern-day playing field:____________________
3. undergo a digital transformation:____________________
4. old guard:____________________
5. drone company:____________________
6. rope-access technicians:____________________
7. IoT devices:____________________
8. capitalize on business intelligence:____________________
9. cloud computing:____________________

10. digital infrastructure:__________

11. streamline manual business operations:__________

Reading B

1. hydrocarbon fuels:__________

2. the digital oilfield:__________

3. the extraction industry:__________

4. visualization platform:__________

5. real-time operations centers:__________

6. oil flows:__________

7. the lack of qualified skilled labor:__________

8. the labor gap:__________

9. staffing costs:__________

10. a development center:__________

11. big data analytics:__________

12. major oilfield services companies:__________

13. reservoir engineering:__________

14. the flow of trapped natural gas:__________

15. rock pores:__________

16. the environmental impact of fracturing:__________

17. fracturing fluid:__________

18. coal gasification projects:__________

19. oil and gas exploration:__________

20. offshore vessels:__________

21. advanced videoconferencing:__________

22. heavy oil:__________

23. light crude(oil):__________

24. diesel fuel:__________

25. organic waste:________________________

26. synthetic crude:________________________

Reading C

1. alternative renewable energy sources:________________________
2. technological advancements:________________________
3. smart drilling:________________________
4. elite runner:________________________
5. well life:________________________
6. operational expertise:________________________
7. fill the void:________________________
8. in the initiation phase:________________________
9. best-in-class technology:________________________
10. the oil and gas sector:________________________
11. the blockchain revolution:________________________
12. the risk of fraud:________________________
13. invalid transactions:________________________
14. financial transactions:________________________
15. claims management:________________________
16. risk management:________________________
17. regulatory compliance:________________________
18. the energy sector:________________________
19. the distributed ledger technology:________________________
20. initial coin offerings:________________________
21. token offerings:________________________
22. green energy trading platforms:________________________
23. the advent of peer-to-peer energy trading :________________________
24. the established norms of energy provision:________________________

25. a pipeline of growth for the future:______

26. run leaner and meaner operations:______

27. unmanned platforms:______

28. generalist manager roles:______

29. niche skill sets:______

30. high capacity wireless platforms:______

31. rig decommissioning:______

32. environmentally-conscious consumers:______

33. carbon-neutral option:______

34. trusted information platforms:______

35. the legacy systems:______

36. a "rip and replace" approach:______

37. digital technologies:______

38. operational efficiency:______

39. digital transformation:______

III Analyze the grammatical structure of the following complex sentences, figure out the meaning of each sentence, and paraphrase them.

1. All organizations existing in traditional verticals can learn how to reignite progress within their companies by assessing the problems that plagued oil and gas, as well as the technological advancements that have the potential to turn things around. (Reading 1, Para. 3)

2. That information is analyzed by clusters of computers, which feed their results to real-time operations centers that adjust oil flows to optimize production and minimize downtimes. (Reading B, Para. 4)

3. This has increased interest in making production and consumption of hydrocarbon fuel less harmful, rather than making alternative fuels commercial. (Reading B, Para. 13)

4. As any elite runner will tell you, the equipment alone doesn't win you the race. Equally, if not

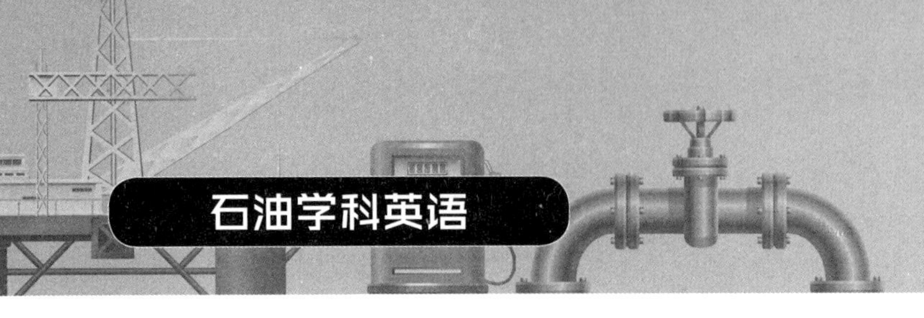

more important, is developing a race plan, road-testing that plan, and developing the intelligence to know exactly when, where and how to hit the gas. (Reading C, Para. 3)

5. As drilling projects grow in ambition, smarter equals faster. By combining integration and intelligence through specialist providers in the initiation phase with best-in-class technology, "smart drilling" promises to give projects the solid footing needed to keep the industry running for decades to come. (Reading C, Para. 6)

6. Blockchain has enormous potential to reduce the risk of fraud, error, and invalid transactions in energy trading, make financial transactions more efficient, facilitate regulatory reporting requirements, and enable interoperability. (Reading C, Para. 8)

7. Finally, the ability to chemically synthesize oil and gas from more types of natural materials will blur the line between renewable and fossil fuels to the extent where it becomes a forgotten issue. (Reading C, Para. 18)

8. The aversion to risk will filter into the organizational culture, with companies looking at how they can run leaner and meaner operations. Having project teams sitting around waiting in the wings for a new assignment will no longer exist. (Reading C, Para. 21)

9. From a technology standpoint, the ideal solution would need to seamlessly connect all systems and hardware platforms across the various fields of operation, integrating exploration, drilling and production facilities, and ultimately delivering useful data and video streams to a central location, allowing the operators to make better and quicker decisions. (Reading C, Para. 24)

Unit 5

Petroleum and Environment

Unit Objectives

Goal 1

Develop a good awareness of the environmental problems associated with the production and use of petroleum through task-based activities.

Goal 2

Understand how these problems impact the environment.

Goal 3

Make an in-depth study of how petroleum industry addresses the environmental issues.

Although petroleum products make life easier, exploring, producing, and using crude oil may have negative effects on the environment. For instance, when petroleum products are burned as fuel, they give off carbon dioxide (CO_2), a greenhouse gas that is linked with global warming. The use of petroleum products also gives off pollutants—carbon monoxide (CO), nitrogen oxides, particulate matter, and unburned hydrocarbons—that pollute the air humans breathe. How to remediate the petroleum-induced environmental degradation is hereby under current research.

Pre-Class Activities

Activity One Search for the Definitions

Search online for the definitions of the following terms or concepts and share your findings with your team members.

1. oil spill:____________________
2. environmental degradation:____________________
3. global warming:____________________
4. freshwater contamination:____________________
5. petroleum-derived contaminants:____________________
6. marine ecosystems:____________________
7. greenhouse gas emissions:____________________
8. exhaust of vehicles:____________________
9. Rig-to-Reefs Program:____________________
10. prudent development:____________________

Activity Two Watch the Videos About Oil Spills

Watch the following two videos and summarize the key points about oil spills in each video.

Video 1: BP Oil Spill Five Years Later: Wildlife Still Suffering

 **Video 2:** How Do We Clean up Oil Spills?

While-Class Activities

Activity One Make a Seminar Discussion

How much do you know about the petroleum-induced environmental issues? Based on the terms above you've learned and your knowledge about this question, make a list of the issues to share with your team members, and finally select one or two issues to present to the whole class.

____

____

____

____

Activity Two Get to Know Petroleum and the Environment

Based on your presentation, read the following article to further comprehend the impacts of petroleum on the environment. You are required to share your comprehension of at least one of the discussed environmental concerns with your team members.

Reading A

Petroleum and the Environment: An Introduction[1]

E. Allison and B. Mandler

❶ When oil and gas were first extracted and used on an industrial scale in the 19th century, they provided significant advantages over existing fuels: They were cleaner, easier to transport, and more versatile than coal and biomass (wood, waste, and whale oil). Diesel and gasoline derived from oil revolutionized the transportation sector. Through developments in chemical engineering, oil and gas also provided the raw materials for a vast range of useful products, from plastics to fertilizers and medicines. By the 20th century, oil and gas had become essential resources for modern life: As both fuel and raw material, the versatility and abundance of oil and gas helped to

---

1. Allison, E., & Mandler, B. (2018, June 01). *Petroleum and the environment: An introduction*. Retrieved from American Geosciences Institute website: https://www.americangeosciences.org/geoscience-currents/petroleum-and-environment-introduction

facilitate unprecedented economic growth and improved human health around the world.

❷ Despite rapid advances in renewable energy technologies, in 2016 oil and gas accounted for two thirds of U.S. energy consumption[1] and over half of all the energy consumed worldwide.[2] Annual U.S. oil and gas production is expected to increase beyond 2040.[3] Developments in policy, technology, and public opinion may change this trend, but oil and gas will likely play a fundamental role in U.S. and global energy production and consumption for much of the 21st century.

Recent Developments

❸ Oil and gas exploration, production, and use have radically changed since the beginning of the 21st century. The use of horizontal drilling with hydraulic fracturing to access previously uneconomic oil and gas deposits led to unprecedented increases in oil and gas production: From 2006 to 2015, U.S. natural gas production increased by 40%,[4] while from 2008 to 2015, U.S. oil production increased by 88%.[5] This growth in production has led to commensurate growth in oil and gas transportation, processing and refining, use in agriculture and manufacturing, and energy exports.[6][7]

❹ However, the recent growth in oil and gas production has increased or renewed some longstanding concerns over their impact on the environment, while also giving rise to some new concerns. Areas of major changes and/or public concerns include:

Hydraulic Fracturing (Fracking)

❺ This technique of fracturing rocks to extract oil and gas has been used since the 1940s, but its combination with horizontal drilling to extract oil and gas from shale led to a surge in hydraulic fracturing starting around 2005. The widespread use of hydraulic fracturing has raised questions about the large amount of water used in the process, which may compete with other fresh water demands in some areas, and has motivated research into alternative fluids. Hydraulic fracturing has also highlighted the issue of groundwater protection, partly due to concerns over the fracturing process itself and partly due to the use of toxic chemicals in some hydraulic fracturing fluids. This adds a new element of concern to a longstanding problem: Old or poorly constructed wells may leak a variety of fluids if the cement or steel portions of the well are compromised, whether they are hydraulically fractured or not. Identifying instances and sources of groundwater contamination is an ongoing challenge for research scientists, regulators, and industry.

Induced Earthquakes

❻ Many human activities can trigger earthquakes, including geothermal energy production, filling up reservoirs behind dams, and groundwater extraction.[8][9] Oil and gas operations can trigger earthquakes through two main processes: underground wastewater injection and hydraulic fracturing. The largest induced earthquakes from oil and gas operations have been caused by the underground injection of large volumes of wastewater extracted along with oil and gas ("produced water").[10] This water is often too salty to release into surface waterbodies so it is instead injected

deep underground, where it can increase the likelihood of earthquakes on existing faults. Hydraulic fracturing very rarely causes noticeable earthquakes, but it has triggered some small but noticeable earthquakes in parts of the United States and Canada.[11] Some states, particularly Oklahoma and Kansas, have observed a decrease in induced seismicity since 2015 due to decreased oil production (and therefore less wastewater in need of disposal) and new regulations constraining wastewater injection volumes and rates.[12]

Land Use

❼ Advances in horizontal drilling mean that wells don't need to be placed directly above a resource, so the location of well sites can be planned to reduce their surface impact, and multiple wells can be drilled in different directions from a single site. However, the boom in horizontal drilling and hydraulic fracturing has led to increased oil and gas activity in many areas, including some areas that had previously had little activity, resulting in increased overall land disturbance in some parts of the country.

Methane Emissions

❽ The surge in U.S. natural gas production has led to natural gas replacing coal as the largest source of electricity in the United States.[13] Burning natural gas releases less carbon dioxide than coal, so in this sense the transition from coal to natural gas has had a positive environmental impact. However, methane, the main component of natural gas, is itself a potent greenhouse gas, so any leaks during natural gas production and distribution will partially offset this benefit. Improved monitoring of methane emissions, targeted repair and replacement of equipment, and potential regulations all play a role in minimizing methane leaks.

Heavy Oil and Oil Sands

❾ Some of the largest oil resources in the world consist of thick, heavy oil that is extracted and processed with specific, energy-intensive techniques. In the United States, California has long produced heavy oil from the Kern River Oil Field outside Bakersfield,[14] but the largest heavy oil producers are Venezuela and Canada. In Canada, a significant proportion of oil production comes from oil sands (also known as "tar sands"), which are a mixture of clay, sand, water, and bitumen (a very thick oil). Canadian oil sand production has increased substantially since 2005. Deeper oil sand deposits are commonly extracted by heating the oil, which thins it so that it can flow up through a well. Shallower deposits are extracted by open-pit mining of the oil sands, which are then processed to remove the oil. Regardless of the production technique, oil sand production is energy-intensive and so results in higher overall emissions of carbon dioxide and combustion-related air pollutants. Open-pit mining of oil sands in particular presents additional environmental concerns, such as risks to air and water quality from dust and waste ponds.

Transportation and Spills

⑩ Increases in oil and gas production and consumption require enhanced transportation infrastructure. About 90% of crude oil and refined products, and essentially all natural gas, are transported through millions of miles of (mostly underground) pipelines. Spills of crude oil and refined products represent less than 0.001% of the total amount transported, but this small percentage amounts to millions of gallons spilled each year.[15] Most spills are small but some can have significant local impacts and require extensive and expensive cleanup efforts.

Offshore Drilling

⑪ Advances in offshore drilling technology have allowed oil and gas to be produced in increasingly deep water. These extreme conditions pose particular technological and environmental challenges. For example, in 2010, a defective well in the Macondo prospect of the Gulf of Mexico caused the largest marine oil spill in history and killed 11 workers on the deepwater horizon drilling rig. Since this spill, regulations and industry practices have changed substantially to reduce the environmental risks of offshore oil and gas production, but many concerns remain.

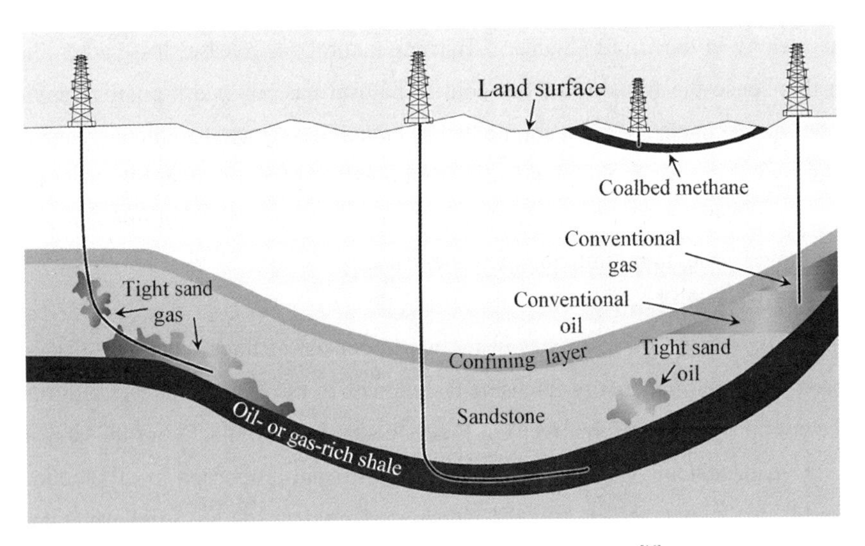

Figure 5-1 Various types of oil and gas deposits[16]

⑫ Recent advances in directional (especially horizontal) drilling and hydraulic fracturing have led to substantial increases in production from shale as well as tight oil and gas sandstone.

A Note on Climate Change

⑬ The combustion of fossil fuels (coal, oil, and natural gas) releases large quantities of carbon dioxide (and other greenhouse gases) into the atmosphere, which has a wide range of environmental impacts. The full extent of these impacts is not yet known, but they include

rising global temperatures, ocean acidification, sea level rise, and a variety of other impacts on weather, natural hazards, agriculture, and more, many of which are likely to increase into the future.[17][18][19] While agriculture and land use change also emit carbon dioxide and other greenhouse gases (especially methane), fossil fuels, especially coal and oil, produce the majority of anthropogenic (human-caused) emissions of greenhouse gases on a global scale.[20]

References

[1] U.S. Energy Information Administration—U.S. Energy Facts Explained.

[2] International Energy Agency—Key World Energy Statistics 2017.

[3] U.S. Energy Information Administration—Annual Energy Outlook 2018.

[4] U.S. Energy Information Administration—Natural Gas Gross Withdrawals and Production.

[5] U.S. Energy Information Administration—Crude Oil Production.

[6] U.S. Pipeline and Hazardous Materials Safety Administration—Data and Statistics Overview.

[7] U.S. Energy Information Administration—Petroleum & Other Liquids: Exports.

[8] National Research Council, "Induced seismicity potential in energy technologies," Washington, D.C.: The National Academies Press, 2013.

[9] G. Foulger et al., "Global review of human-induced earthquakes," *Earth-Sci. Rev.*, vol. 178, pp. 438–514, 2017.

[10] U.S. Geological Survey—Induced Earthquakes: Myths and Misconceptions.

[11] G. M. Atkinson et al., "Hydraulic fracturing and seismicity in the Western Canada sedimentary basin." *Seismol. Res. Lett.*, vol. 87, no. 3, pp. 631–647, 2016.

[12] American Geosciences Institute—State Responses to Induced Earthquakes.

[13] U.S. Energy Information Administration—What Is U.S. Electricity Generation by Energy Source?

[14] S. Weeden, "Steamflooding keeps California field producing 117 years later," *E & P Mag.*, Apr. 1, 2016.

[15] U.S. Pipeline and Hazardous Materials Safety Administration—Incident Statistics.

[16] U.S. Environmental Protection Agency—Assessment of the Potential Impacts of Hydraulic Fracturing for Oil and Gas on Drinking Water Resources.

[17] National Oceanic and Atmospheric Administration—Climate Change Impacts.

[18] NASA—The Consequences of Climate Change.

[19] Intergovernmental Panel on Climate Change—Climate Change 2013: The Physical Science Basis.

[20] U.S. Environmental Protection Agency—Global Greenhouse Gas Emissions Data.

After-Class Activities

Activity One Read, Review, and Report

First read the following chart to review what you've learned in Reading A and report orally about what you can summarize to the whole class.

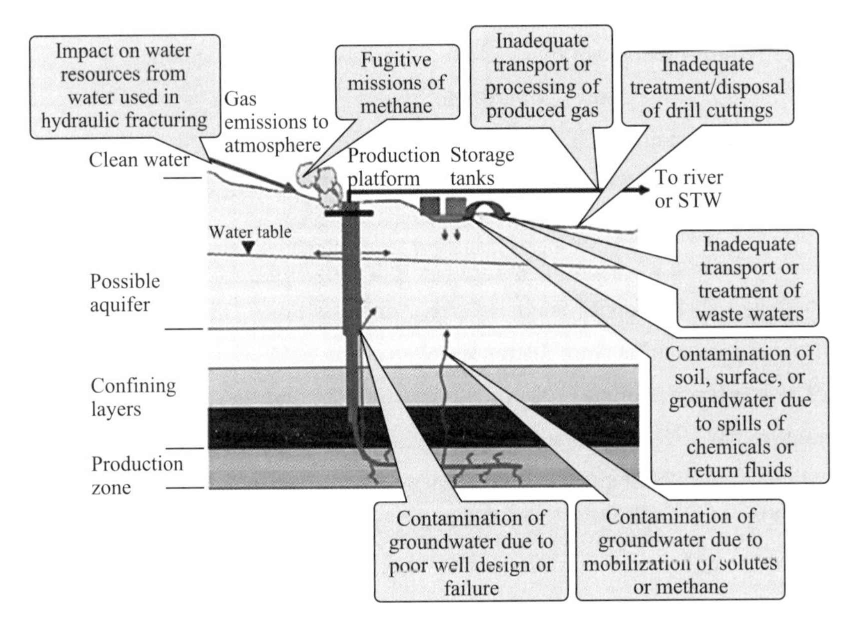

Figure 5-2 Environmental risks of shale gas extraction
(Source: Environment Agency.)

Activity Two Watch the Video About the Environmental Impact

Watch the following video and make a summary about the impact discussed.

 **Video:** The Environmental Impact of Hydraulic Fracking

Activity Three Practice Summary Writing

Topic sentences can help you a lot in writing a summary. The following is an excerpt of the article, giving a review of the environmental impacts of petroleum. Read it carefully and then summarize the key points with not more than 150 words.

Reading B

Petroleum and the Environment[1]

Onyenekenwa Cyprian Eneh

❶ Even though petroleum products make life easier by their use in fueling airplanes, cars, trucks, cooking stoves, and other applications in combustion engines, as well as in heating our homes, using them can cause environmental problems, like air and water pollution. When petroleum products are burned as fuel, they give off carbon dioxide (CO_2), a greenhouse gas that is linked with global warming. The use of petroleum products also gives off pollutants—carbon monoxide (CO), nitrogen oxides, particulate matter, and unburned hydrocarbons—that pollute the air humans breathe. Since a lot of air pollution comes from cars and trucks, many environmental laws have been aimed at changing the make-up of gasoline and diesel fuel so that they produce fewer emissions. These "reformulated fuels" are much cleaner burning than gasoline and diesel fuel were in 1990. In the next few years, the amount of sulphur contained in gasoline and diesel fuel will be reduced dramatically so that they can be used with new, less-polluting again technology.[1]

❷ Exploring and drilling for oil may disturb land and ocean habitats. A study on assessment of the impact of oil exploration activities on agriculture and natural resources in the Niger Delta Region of Nigeria showed that oil exploration activities adversely impacted specifically on soil/land resources, aquatic life/fisheries, water resources, crops, livestock, and forests/vegetation. Oil spills have degraded most agricultural lands, reduced the availability of fish and fish products, and caused the pollution of surface and ground water resources, destruction of arable and tree crops, and death of farm animals in the region as a result of toxic materials in the soil and polluted water. Oil exploration activities have also resulted in the disappearance of some forest vegetation and animal species, including primates, fish, turtles, and birds. The ultimate result of these impacts is a drastic reduction in farm productivity and animal farm income.[2][3][4][5][6]

❸ New technologies have greatly reduced the number and size of areas disturbed by drilling, sometimes called "footprints". Satellites, global positioning systems, remote sensing devices, and 3D and 4D seismic technologies, make it possible to discover oil reserves while drilling fewer

1. Eneh, O. C. (2011). A review on petroleum: Source, uses, processing, products, and the environment. *Journal of Applied Sciences, 11*: 2084–2091. DOI: 10.3923/jas.2011.2084.2091

wells. Again, the use of horizontal and directional drilling makes it possible for a single well to produce oil from much bigger areas. Today's production footprints are only about one fourth the size of those of 30 years ago, due to the development of movable drilling rigs and smaller "slimhole" drilling rigs. When the oil in a well is gone, the well must be plugged below ground, although its soil fertility is gone. As part of the "Rig-to-Reefs" program, some old offshore rigs are toppled and left on the sea floor to become artificial reefs that attract fish and other marine life. Within six months to a year after a rig is toppled, it becomes covered with barnacles, coral, sponges, clams, and other sea creatures.[1]

❹ If oil is spilled into rivers or oceans, it can harm wildlife. Oil spills can come from natural oil seeps from the ocean floor, ships that crash, or leaks that happen when petroleum products are used on land, such as the gasoline that sometimes drips onto the ground when people are filling their gas tanks, motor oil that gets thrown away after an oil change, or fuel that escapes from a leaky storage tank. When it rains, the spilled products get washed into the gutter and eventually go to rivers and the ocean. Another way that oil sometimes gets into water is when fuel is leaked from motorboats and jet skis.[1]

❺ When a leak in a storage tank or pipeline occurs, petroleum products can also get into the ground. In some places where gasoline has leaked from storage tanks, one of the gasoline ingredients, called Methyl Tertiary Butyl Ether (MTBE), has made its way into local water supplies. Since MTBE makes water taste bad and many people are worried about drinking it, a number of states are banning the use of MTBE in gasoline and the refining industry is voluntarily moving away from using it when blending reformulated gasoline. To prevent leaks from underground storage tanks, all buried tanks are supposed to be replaced by tanks with a double lining.[1]

References

[1] D. S. Ugwu, "Effects of oil exploration on agriculture and natural resources in the niger delta region of Nigeria," *Sustainable Human Dev. Rev*., vol. 1, pp. 117–130, 2009.

[2] A. I. Al-Turki, "Assessment of effluent quality of tertiary wastewater treatment plant at buraidah city and its reuse in irrigation," *J. Applied Sci*., vol. 10, pp. 1723–1731, 2010.

[3] O. C. Eneh, "Effects of water and sanitation crisis on infants and under-five children in Africa, " *J. Environ. Sci. Technol*., vol. 4, pp. 103–111, 2011.

[4] O. C. Eneh, "Managing Nigeria's environment: The unresolved issues," *J. Environ. Sci. Technol*., vol. 4, pp. 250–263, 2011.

[5] M. A. S. Tabieh, and A. Al-Horani, "An economic analysis of water status in Jordan," *J.*

Applied Sci., vol. 10, pp.1695–1704, 2010.

[6] U.S. Department of Energy Office of Fossil Energy—Environmental Benefits of Advanced Oil and Gas Exploration and Production Technology. Retrieved from http://fossil.energy.gov/news/techlines/1999/tl_envrpt.html

Activity Four Address Petroleum-Induced Environmental Issues

Petroleum industry has been going all out to ameliorate the environmental risks for sustainability. The following article reviews what oil companies have done to protect the environment while meeting the ever-increasing energy needs. Summarize the chief acts they have performed, and report orally your findings to the whole class.

Reading C

Oil and the Environment—What Are Oil Companies Doing to Clean up Their Act[1]

❶ The world continues to thirst for oil with an ever increasing fervor, yet simultaneously struggles to fully grasp and appreciate the obstacles encountered by those who bring that oil to their local pumps.

❷ There has always been a necessary trade-off when it comes to technology and industrial advancement, as in order to experience the many benefits modern society offers us, we have to agree to give up some portion of nature and accept a certain amount of environmental damage.

❸ Logic dictates that "we can't have our cake and eat it too", but recently the oil industry has been seeking ways in which they can continue to bring us the oil that we thrive upon and still protect the environment that it is derived from.

❹ Fact 1: Our world continues to demand a consistent and increasingly larger flow of oil to supply its constantly growing needs for transportation and energy as second and third world nations quickly progress into what may soon become an ever expanding, "equal playing field" of first world civilization.

❺ Fact 2: Oil is a very messy business. There are horrendous calamities, such as oil spills,

---

1. Adapted from *Oil and the environment—what are oil companies doing to clean up their act.* (2009, December 04). https://oilprice.com/The-Environment/Global-Warming/Oil-And-The-Environment-What-Are-Oil-Companies-Doing-To-Clean-Up-Their-Act.html

pipeline ruptures, fires, explosions as well as rigs breaking, ships sinking, trains derailing, and trucks jack-knifing or crashing. Toss the polluting potential involved in the refining process as well as intentional havoc created by acts of sabotage and terrorism, and you have yourself one powerful powder keg to protect.

❻ "Necessity is the mother of invention!" Today's oil industry realizes that they have a dark cloud of negative press and public perception hovering ominously over their head and are wisely refocusing their attention on effectively cleaning up their act and reputation, while learning to better preserve the very environment which they rely upon to draw their profits and sustenance.

❼ Meanwhile, those once "die hard" environmentalists now comprehend that it's actually in everyone's best interests if they now work cooperatively alongside the oil companies, helping them to succeed in a way that takes far less toll on our planet and the multiple millions of people inhabiting it.

❽ Here's what this newly forged team of former foes finds itself up against:

❾ Every single year, millions of gallons of oil are released into our environment due to technological and mechanical breakdown, human error or carelessness, as well as natural disasters such as earthquakes, hurricanes, fires, and explosions. Not to mention catastrophes created by certain state sponsored acts of aggression or warfare and of course, the blatant devastation of terrorism.

❿ Oil companies, working alongside environmental organizations, scientists, biologists, and engineers have developed numcrous solutions for spills occurrıng both on land and in the water. Water spills can now effectively be cleaned by presses that involve straining and draining while containing the oil slick, using "floating booms" to corral the oil while skimmers and vacuum pumps cleanse the water and reclaim large percentages of the spilled oil.

⓫ Another "eco-friendly" oil spill management method for both water and land spills is called "bio-remediation". It's a technique that uses living organisms such as bacteria and fungi to degrade, break down, and in some cases, actually eat the oil as it safely cleanses the spill without hurting the environment. Meanwhile, serious upgrading of the technology now being used for the drilling and refining of oil is cutting previous pollution levels down tremendously, as the oil industry increases profits by processing more usable oil while polluting a lot less.

⓬ Oil companies are now investing billions of dollars in socially responsible programs and are quickly becoming one of the largest supporters of environmentally friendly programs worldwide. Oil companies have been already largely responsible for many of the major advances in medicine, pharmacology, and worldwide health care infrastructure, but now they are some of the largest supporters of research dedicated to promoting renewable energy sources.

⓭ British Petroleum is a leader in solar, wind, and hydrogen power programs. Meanwhile,

Exxon Mobil is partnering with Stanford University on its Global Climate and Energy Project! Oil companies are some of the largest contributors actively partnering in Third World developments, investing in their future by donating billions of dollars for the building of schools, hospitals, libraries, and other much needed infrastructure, getting countless previously unemployed workers off of the streets and feeding families that otherwise would starve to death.

⑭ Exxon, after the media blood bath it suffered over the Valdez catastrophe, has drastically improved its reputation worldwide, but particularly with the environmental groups, by paying for all of the damages and working diligently to promote resurgence of wildlife in Prince William Sound, as well as funding environmental education to prevent and protect against future spills. They also recently donated millions to the Save the Tiger Fund, the World Wildlife Fund, and the Wildlife Conservation Society.

⑮ In 2008, at Chevron's annual shareholders meeting, the priority items of significance placed on the agenda by the stockholders included: environmental justice, peace, and human rights worldwide. Meanwhile, Innovest Strategic Value Advisors, an investment advisory firm dedicated to helping concerned investors find environmental leaders to invest in, proudly recommended both Shell and BP for their superior environmental management programs and commitment to sustainable development.

⑯ While Shell, BP, and Amoco were rated particularly high for their work on renewable energy, corporate social responsibility, and research regarding climate change, they gave strong referrals to Exxon, for their efforts to develop a now highly respectable environmental management framework directed at reducing environmental impact while improving performance. They also glowingly recognized Texaco for their expertise in waste "gasification" that they are not only using for their own benefits, but also making available to competitors.

⑰ The oil industry has recently done a dramatic reversal in its methodology of doing business and this new commitment to the environment and people who live within it, including themselves, is going a long way in bolstering their global reputation, while making them a whole new set of friends and allies, many of whom were once their biggest and most dedicated detractors! This truly is excellent news for every living being on the planet, as well as the planet itself!

Integrated Exercises

I Read the following academic words, and check whether you can use them appropriately. For those you can not, look up in a dictionary or search online about their contextual use. Write down notes to strengthen your memory.

Reading A

1. transportation
2. abundance
3. revolutionize
4. fertilizer
5. unprecedented
6. exploration
7. radically
8. deposit
9. refine
10. longstanding
11. surge
12. motivate
13. leak
14. cement
15. compromise
16. contamination
17. regulator
18. trigger
19. ongoing
20. injection
21. induce
22. constrain
23. multiple
24. disturbance
25. potent
26. minimize
27. proportion
28. substantially
29. energy-intensive
30. enhance
31. cleanup
32. pose
33. defective
34. prospect
35. combustion
36. emit

Reading B

1. formulate
2. habitat
3. assessment
4. adversely
5. vegetation
6. availability
7. destruction
8. arable
9. toxic
10. productivity

11. drastic
12. fertility
13. marine
14. plug
15. toppled
16. seep
17. leaky
18. ban
19. ingredient

Reading C

1. fervor
2. obstacle
3. encounter
4. pump
5. thrive
6. calamity
7. messy
8. rupture
9. toss
10. havoc
11. sabotage
12. perception
13. ominously
14. sustenance
15. forge
16. foe
17. warfare
18. explosion
19. catastrophe
20. sponsor
21. aggression
22. blatant
23. drain
24. contain
25. cleanse
26. devastation
27. numerous
28. alongside
29. press
30. strain
31. skimmer
32. vacuum
33. reclaim
34. degrade
35. bio-remediation
36. tremendously
37. pharmacology
38. dedicate
39. solar
40. donate
41. drastically
42. diligently
43. conservation
44. starve
45. commitment
46. expertise
47. detractor

II Decide on the contextual meaning of the following terms and expressions.

Reading A

1. transportation sector:________________
2. raw material:________________
3. energy consumption:________________
4. horizontal drilling:________________
5. public concern:________________
6. the issue of groundwater protection:________________
7. hydraulic fracturing:________________
8. geothermal energy:________________
9. groundwater extraction:________________
10. underground wastewater injection:________________
11. produced water:________________
12. surface waterbody:________________
13. well site:________________
14. a potent greenhouse gas:________________
15. oil sand:________________
16. tar sand:________________
17. deeper oil sand deposit:________________
18. open-pit mining:________________
19. combustion-related air pollutant:________________
20. offshore drilling technology:________________
21. enhanced transportation infrastructure:________________
22. environmental risks:________________
23. ocean acidification:________________
24. sea level rise:________________
25. natural hazard:________________

Reading B

1. petroleum product:__________
2. combustion engine:__________
3. particulate matter:__________
4. oil exploration activities:__________
5. forest vegetation and animal species:__________
6. farm productivity:__________
7. animal farm income:__________
8. global positioning system:__________
9. remote sensing devices:__________
10. 3D and 4D seismic technologies:__________
11. directional drilling:__________
12. movable drilling rigs:__________
13. offshore rigs:__________
14. artificial reefs:__________
15. marine life:__________
16. sea creatures:__________
17. gas tanks:__________
18. motor oil:__________
19. storage tank:__________
20. the refining industry:__________

Reading C

1. equal playing field:__________
2. pipeline rupture:__________
3. rig breaking:__________
4. train derailing:__________

5. powder keg:______

6. acts of sabotage and terrorism:______

7. technological and mechanical breakdown:______

8. oil slick:______

9. floating boom:______

10. oil spill management method:______

11. environmentally friendly programs:______

12. health care infrastructure:______

13. hydrogen power:______

14. environmental groups:______

15. an investment advisory firm:______

16. environmental management programs:______

17. sustainable development:______

18. corporate social responsibility:______

III Analyze the grammatical structure of the following complex sentences, figure out the meaning of each sentence, and paraphrase them.

1. The use of horizontal drilling with hydraulic fracturing to access previously uneconomic oil and gas deposits led to unprecedented increases in oil and gas production: From 2006 to 2015, U.S. natural gas production increased by 40%, while from 2008 to 2015, U.S. oil production increased by 88%. (Reading A, Para. 3)

2. This adds a new element of concern to a longstanding problem: Old or poorly constructed wells may leak a variety of fluids if the cement or steel portions of the well are compromised, whether they are hydraulically fractured or not. (Reading A, Para. 5)

3. However, the boom in horizontal drilling and hydraulic fracturing has led to increased oil and gas activity in many areas, including some areas that had previously had little activity, resulting in increased overall land disturbance in some parts of the country. (Reading A, Para. 7)

4. Oil spills can come from natural oil seeps from the ocean floor, ships that crash, or leaks that happen when petroleum products are used on land, such as the gasoline that sometimes drips onto the ground when people are filling their gas tanks, motor oil that gets thrown away after an oil change, or fuel that escapes from a leaky storage tank. (Reading B, Para. 4)

5. Since MTBE makes water taste bad and many people are worried about drinking it, a number of states are banning the use of MTBE in gasoline and the refining industry is voluntarily moving away from using it when blending reformulated gasoline. (Reading B, Para. 5)

6. The world continues to thirst for oil with in an ever increasing fervor, yet simultaneously struggles to fully grasp and appreciate the obstacles encountered by those who bring that oil to their local pumps. (Reading C, Para. 1)

7. Logic dictates that "we can't have our cake and eat it too", but recently the oil industry has been seeking ways in which they can continue to bring us the oil that we thrive upon and still protect the environment that it is derived from. (Reading C, Para. 3)

8. Toss the polluting potential involved in the refining process as well as intentional havoc created by acts of sabotage and terrorism, and you have yourself one powerful powder keg to protect. (Reading C, Para. 5)

9. Today's oil industry realizes that they have a dark cloud of negative press and public perception hovering ominously over their head and are wisely refocusing their attention on effectively cleaning up their act and reputation, while learning to better preserve the very environment that they rely upon to draw their profits and sustenance. (Reading C, Para. 6)

10. Water spills can now effectively be cleaned by presses that involve straining and draining while containing the oil slick, using "floating booms" to corral the oil while skimmers and vacuum pumps cleanse the water and reclaim large percentages of the spilled oil. (Reading C, Para. 10)

Unit 6

Petroleum and Education

Unit Objectives

Goal 1

Get informed of the knowledge, skills, and qualities required by employers in the oil and gas industry through task-based activities.

Goal 2

Get to know how education can be positioned to provide employers in the oil and gas industry with workers who possess appropriate knowledge and skills to meet the needs.

Goal 3

Get a glimpse of the challenges education will be undertaking to cater to the ever-changing demands of the oil and gas industry.

Currently and for the foreseeable future, oil and gas will continue to be the world's primary energy feedstock. This means that petroleum assets are as critical as they have been during the last century to meet the energy demands globally. If the industry continues its mission in providing oil and natural gas resources to the world communities, it will need the services of appropriately skilled professionals. Oil industry hires many technical professionals. At the core, however, there are unique petroleum engineering concepts and domain knowledge that define the industry and its technology base. It probably fits better in describing the sciences and techniques as a broader base of subsurface engineering rather than energy engineering. This particular domain expertise can be gained in depth via formal university degrees or through continuing education courses and augmented with the job training.

Pre-Class Activities

Activity One Search for the Definitions

Search online for the definitions of the following terms or concepts and share your findings with your team members.

1. petroleum engineer:____________________
2. engineering technician and technologist:____________________
3. oil and gas workforce:____________________
4. engineering specialty:____________________
5. entry-level jobs/employment:____________________
6. in-house/on-the-job training:____________________
7. industry internships or placements:____________________
8. industry-academia partnership:____________________
9. cross-cutting skills or knowledge:____________________
10. critical thinking:____________________
11. time management:____________________

Activity Two Watch the Videos About Top Qualities

Watch the following two videos about the qualifications oil and gas companies are looking forward to and summarize the key points for a team presentation.

 **Video 1:** How to Become a Drilling Engineer: Education and Career Roadmap

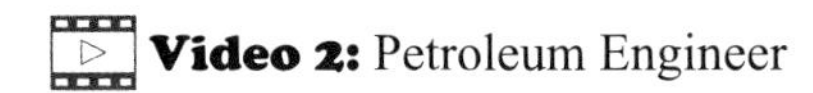
Video 2: Petroleum Engineer

While-Class Activities

Activity One Analyze the Figures

The following figures provide you with several surveys published in 2017 about the aspects the American oil and gas employers are looking for. The surveys were undertaken jointly by RAND Labor and Population and RAND Education. Read the figures to analyze what these surveys indicate, write down your own findings, and make a class presentation.

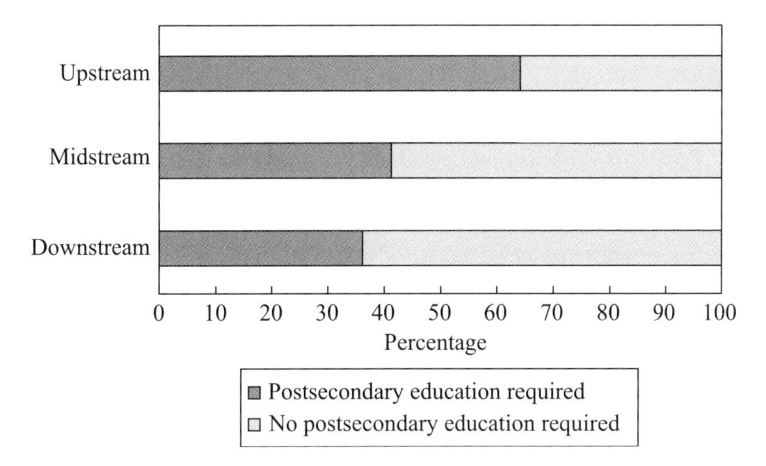

Figure 6-1 Postsecondary educational requirements of projected job openings in the oil and natural gas industry from 2014 through 2024, by segment of the industry (Source: U.S. Bureau of Labor Statistics 2016.)

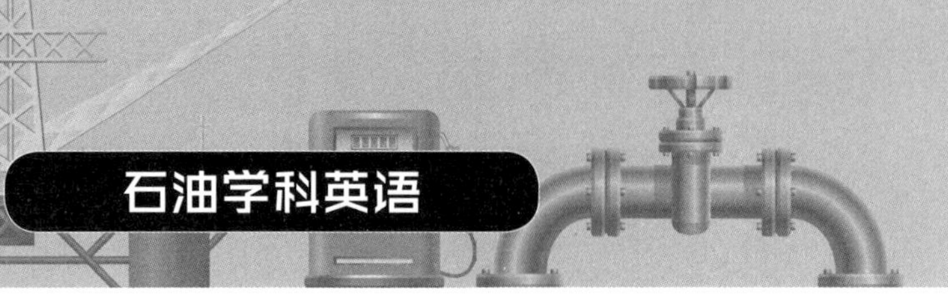

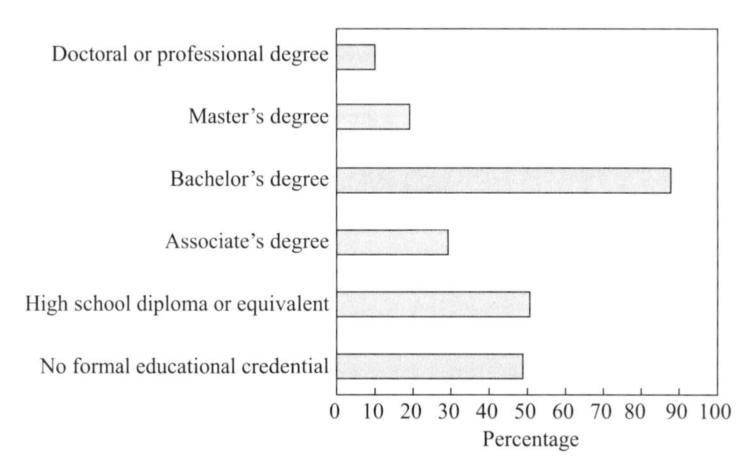

Note: N = 67. Results are weighted to generalize to all oil and natural gas employers in Pennsylvania and West Virginia. Responses do not add to 100% because respondents were allowed to mark multiple high-priority occupations.

Figure 6-2 Percentage of employers with high-priority occupations that require specific levels of education

(Source: RAND SHALE Survey of Employers 2016–2017 and O* NET Resource Center.)

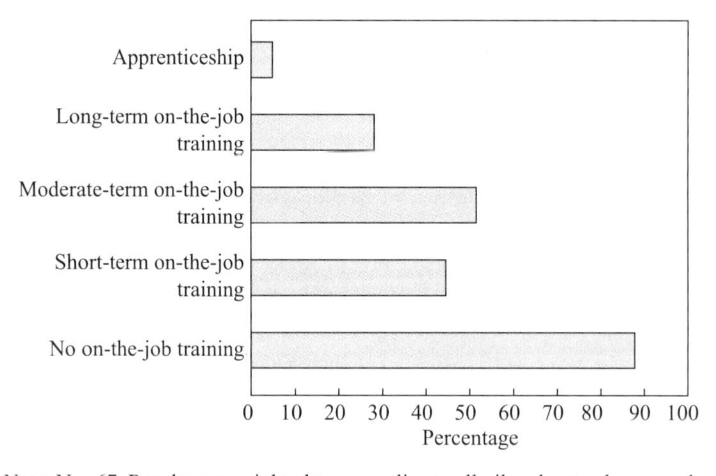

Note: N = 67. Results are weighted to generalize to all oil and natural gas employers in Pennsylvania and West Virginia. Responses do not add to 100% because respondents were allowed to mark multiple high-priority occupations.

Figure 6-3 Percentage of employers with high-priority occupations that require specific levels of on-the-job training needed to obtain competency in high-priority occupations

(Source: RAND SHALE Survey of Employers 2016–2017 and U.S. Department of Labor Employment and Training Administration O* NET Survey.)

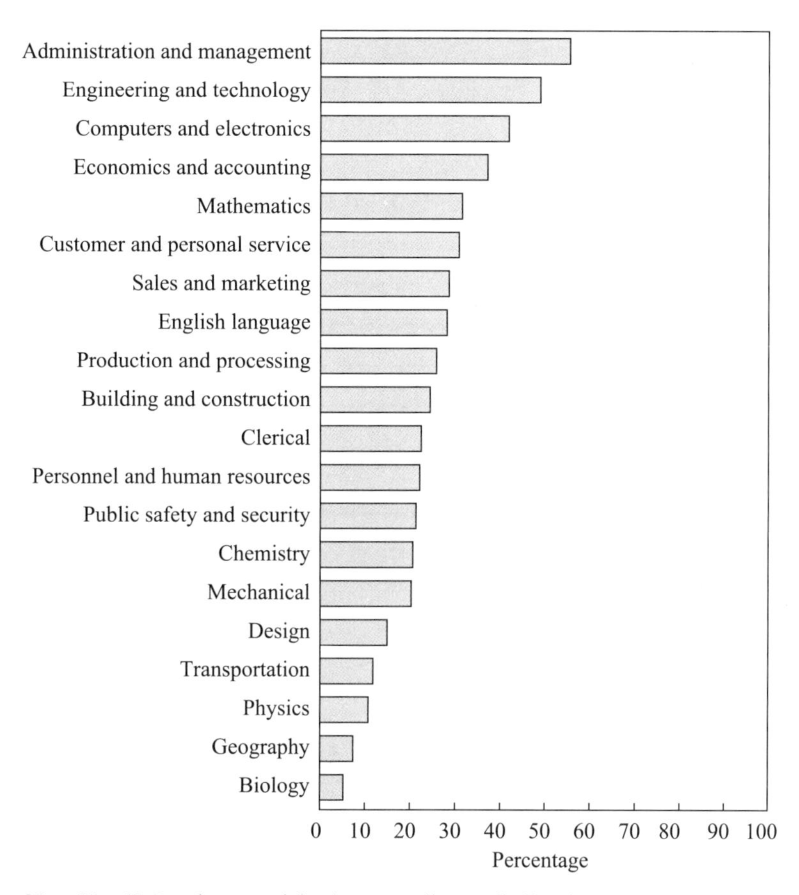

Note: N = 67. Results are weighted to generalize to all oil and natural gas employers in Pennsylvania and West Virginia. Responses do not add to 100% because respondents were allowed to mark multiple high-priority occupations.

Figure 6-4 Percentage of employers reporting types of knowledge needed for high-priority occupations

(Source: RAND SHALE Survey of Employers 2016–2017.)

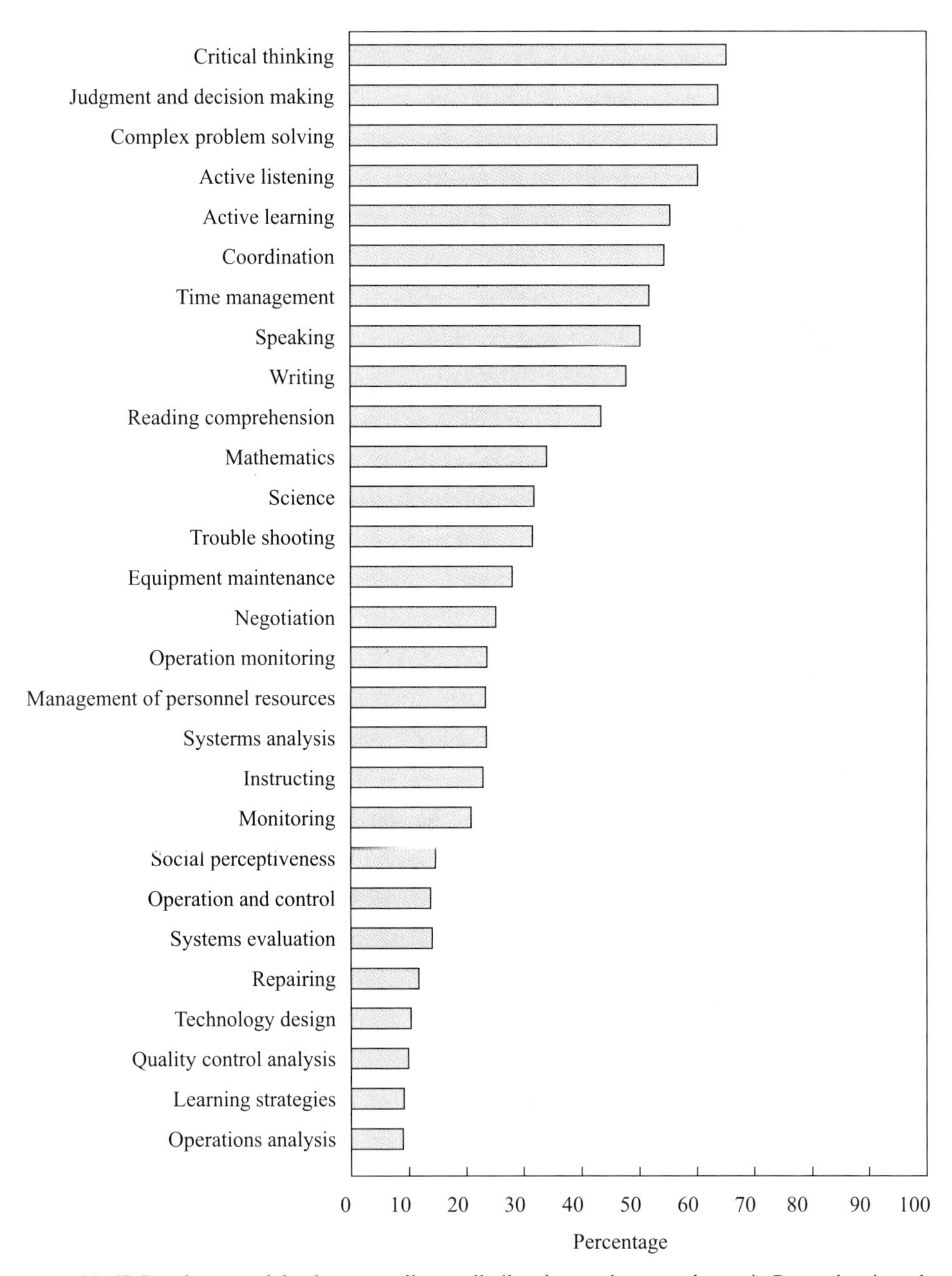

Note: N=67. Results are weighted to generalize to all oil and natural gas employers in Pennsylvania and West Virginia. Responses do not add to 100% because respondents were allowed to mark multiple high-priority occupations.

Figure 6-5 Percentage of employers reporting specific skills sought for high-priority occupants

(Source: RAND SHALE Survey of Employers 2016–2017.)

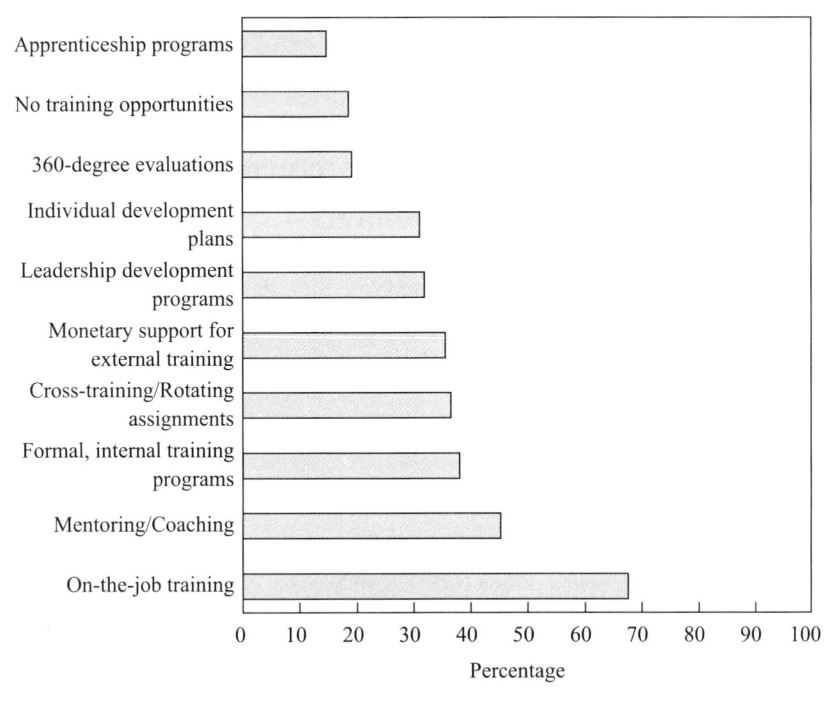

Note: N = 67. Results are weighted to generalize to all oil and natural gas employers in Pennsylvania and West Virginia. Responses do not add to 100% because respondents were allowed to mark multiple high-priority occupations.

Figure 6-6 Percentage of employers reporting training provided to employees to meet skills needs in high-priority occupations

(Source: RAND SHALE Survey of Employers 2016–2017.)

Activity Two Make a Seminar Discussion

Based on what you've learned from the videos and the figures above, hold a discussion with your team members about the essential qualities imposed by the petroleum industry and the suggestions on what education can do to meet the needs. Finally, focus on Reading A to know the education requirements for becoming a petroleum engineer. Summarize the key points listed in the reading.

What Are the Education Requirements to Become a Petroleum Engineer?[1]

❶ A petroleum engineer searches for gas and oil, then designs new ways to extract such materials from the Earth, and makes them into usable fuel for consumers. The majority of petroleum engineers are only required to hold a bachelor's degree in their engineering specialty for entry-level jobs, but some may pursue a graduate degree for advanced research positions.

High School Education

❷ High school students who want to become petroleum engineers should take advanced courses in calculus, chemistry, earth science, computer science, English, and physics. Students should obtain good grades in order to get into a well-known college and enter its engineering program. High school students should also prepare for this kind of career by participating in an engineering club, such as the Junior Engineering Technical Society or a geology club.

Table 6-1 Important facts about petroleum engineering careers

Median salary (2014)	$130,050
Job outlook (2014–2024)	10%
Key skills	Analytical skills, creativity, math skills
Work environment	Offices, research laboratories, drilling sites

(Source: U.S. Bureau of Labor Statistics.)

Postsecondary Education

❸ Aspiring petroleum engineers should attend a college with an engineering program approved by ABET (the Accreditation Board for Engineering and Technology). Students should have excellent grades in math and science in order to be accepted into the college's undergraduate engineering program. Students can complete their bachelor's degree in about four or five years. During their undergraduate studies, students may explore other specialties, such as chemical engineering, civil engineering, materials engineering, geology, and chemistry. Students can also pursue a graduate degree if they wish to obtain faculty positions or advanced research positions in their field, but it is not required for entry-level employment.

1. Adapted from *What are the education requirements to become a petroleum engineer?* (unknown publication date). Retrieved from https://learn.org/articles/What_are_the_Educational_Requirements_to_Become_a_Petroleum_Engineer.html

Required Courses

❹ Petroleum engineers need a well-rounded background in general engineering concepts that can appeal to employers in different engineering fields. This background is provided by coursework in chemistry, geology, and engineering systems. Petroleum engineering programs also include coursework in well design and petrophysics, as well as a drilling and production engineering lab.

Licensing and Certification

❺ Engineers who offer their services directly to the public in the United States must have a license, and they are called professional engineers (PEs). If they don't directly serve the public, they don't need a license to work. Petroleum engineers may consider pursuing certification. Certification may facilitate job advancement to managerial positions and demonstrate competency. The Society of Petroleum Engineers is one of the many professional organizations that offer certification exams, professional conduct guidelines, seminars, career tools, and networking events.

Activity Three Get to Know the Essential Qualities

What are the essential qualities petroleum engineers need to have? Based on your understanding through the reading and activities above, read the following article which takes the Canadian oil and gas companies as an example to illustrate the six skills or qualities imposed on petroleum employees. Interpret orally what you've learned about each skill and share with your team members. Then based on what you've comprehended, give your own ideas about what education can do to fulfill the needs.

Reading B

Six Skills Oil and Gas Companies Are Seeking in Employees[1]

Joel Walker

CDI Corporation

❶ The well-worn expression "The only thing that is constant is change" from our old Greek friend Heraclitus, could not be more true or pronounced than it is today. Over the last half century, global energy development and exploration has gone through considerable change. Not only have

1. Walker, J. (2016, April 15). *Six skills oil and gas companies are seeking in employees*. Retrieved from http://www.oilgasmonitor.com/operating-companies-oilfield-service-companies-maintaining-patent-rights-2/

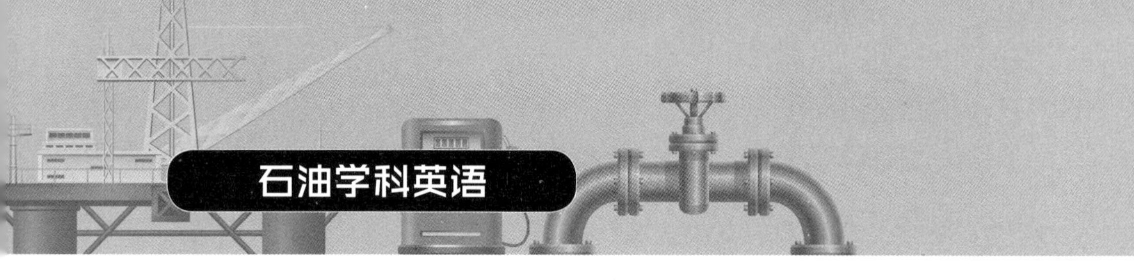

there been vast advances in technologies, materials, and processes, we have seen extreme volatility in commodity markets, increased environmental regulation, and a strong push towards better consultation with First Nations stakeholders.

❷ For Canadian energy companies to successfully meet the challenge of this dynamic landscape, not only will certain skills be more in demand than others, how and at what point in people's careers they acquire those skills is shifting.

❸ As a program manager in the professional services field, I get to experience the frontlines of skill assessment, upgrading, and compliance. We are fortunate that our clients are often able to clearly articulate to us exactly the skill sets that they require. Through this search and selection process, a picture emerges where we can see what the six hottest skills in demand today.

Soft Skills

❹ Soft skills refer to people management, interpersonal skills, communication, and teamwork. We all learned the basics of these skills in the schoolyard, but ask yourself, "What would the outcome of my last project be?", if you worked with a team that was not skilled at these things. Then ask yourself, "How much better could the project have been with the practice of exceptional soft skills?" We consistently find that resources who are considered to be adept with soft skills and people management are requested repeatedly and have better overall project outcomes.

❺ The increasing diversity of culture, language, age, and background of those in our workforce requires that managers and co-workers have a greater depth of understanding and utilization of all the soft skills. A wide range of courses in leadership, dispute resolution, and effective communication are accessible through colleges, technical schools, and increasingly online.

Transferable Skills

❻ The key skills required in oil and gas industry are not that different from those required in most Canadian industries. Growing opportunities exist for those that can demonstrate how their previous skills and expertise can be used to create new solutions within traditional oil and gas models.

❼ One of the sought after transferable skill sets required for work that we perform in the construction management and inspection business is previous military service. Many of the leadership, safety, and project planning skills that form the basis of military life are very well-suited to oil and gas field projects. With some brief research on a target company, candidates are able to present their depth of skill in terms that are attractive to hiring managers looking to find solutions.

Ability to Change

❽ New procedures, certifications, and approaches are being developed at a rapid pace. We often don't get the luxury of simply learning as we go. Market changes can be upon us quickly

and our ability to "dance the new dance" is critical to success and sustainability. We currently find ourselves in a very challenging energy economy; however as little as 18 months ago, the market was quite different. The capacity to effectively pivot from one mode or approach to an entirely different one, means a huge reduction in rework and an increase in quality.

❾ Previous exposure to change management principles, taking on new projects that may lie outside your traditional role and enrolling in continuing education courses, are all ways to remain mentally flexible and display an appetite for the dynamic world of oil and gas.

Data Analytics (Discovery and Communication of Information Patterns)

❿ We are inundated with a constant flow of information. The ability to collect, understand, and analyze this mass of information is not only critical to success, but is required at every role and level of an organization. While the title appears daunting, one does not need to be a Nobel winning statistician to develop some good skills around measuring and tracking some basic metrics. Asking "why?" "what if?" and "how come?" are good starting points that could lead to major process or system improvements.

⓫ The continuing education departments of most Canadian universities and colleges offer introductory and advanced courses in data analysis. Whether you work in an upstream or downstream portion of the business, greater understanding of data relationships and patterns is a highly valuable tool.

Technical Skills

⓬ Every job requires a certain level and set of skills, but how does one pick the skill that is most in demand? Apologies in advance, this is a trick question! What is more important than any particular skill, and critical to the oil and gas industry is constant skill improvement. Upgrading previous skills, adding new ones, and a willingness to learn are all part of keeping skills sharp. From a hiring manager's perspective, continued education communicates key information about a candidate's relevance in the market, competency to deliver on expectations, and the ability to change with the technical needs of the business.

⓭ Training certifications and compliance are forefront in ensuring required skills are present and in good working order. Regulatory oversight will only continue to grow as energy projects expand in scope, size, and need for public support.

⓮ Leading the way in contractor compliance, CDI has created a custom, cloud-based software that stores, tracks, and automatically reminds contractors of upcoming certificate expiry dates. Not only does this provide real time, auditable compliance reporting for our clients, it also provides our contractors the tools required to ensure compliance at all times throughout a project.

Innovation and Creative Solutions

⓯ Some of the issues being faced today are entirely new for the industry and require a fresh perspective to create different outcomes. Traditional processes and methods are not suited to many of the changes and challenges that arise with the introduction of new technologies, expanded regulations, and new modes of engagement with stakeholders.

⑯ While the amount of experience and years of education build a basket of knowledge, these two factors can also be the largest impediment to creative solutions. The ability to assess obstacles through the lens of imagination, risk-taking and non-traditional assessment models will provide the kind of new, useful, and feasible solutions required.

⑰ Marshall McLuhan, the famous Canadian media theorist captured this need for innovation with his statement, "Our Age of Anxiety is, in great part, the result of trying to do today's job with yesterday's tools and yesterday's concepts."

⑱ Thinking back to our Greek friend Heraclitus, we are reminded that change is constant and therefore the tools, skills, and modes of thinking required for success will forever change. New modes of education, including lifelong learning, adapting quickly to change, and transferring skills from one industry to another, are all key parts of the exciting opportunities in the Canadian oil and gas business.

After-Class Activities

Activity One Watch the Video About Petroleum Education

Watch the following video and make a summary about what makes the top petroleum engineering schools stand out.

 **Video:** List of Top Petroleum Engineering Schools and Colleges

Activity Two Practice Summary Writing

What should colleges and universities do to foster the above-mentioned qualities or skills? The following article may be a good source of reference about what colleges and universities should drive at to meet the ever-increasing needs in oil and gas industry. Read it carefully and write a summary with not more than 150 words.

Reading C

Looking Ahead: Challenges for Petroleum Engineering Education[1]

Cunha, J. C. and Cunha, L. B.

Introduction

❶ "Petroleum engineers make the world run." This proud quote, extracted from a petroleum engineering society's website, indicates how highly we regard our career and, at the same time, indicates how important it should be in the educational process that prepares the next generation of engineers to fulfill the industry needs.

❷ Petroleum engineering, as a formal academic course, is about to complete its first century[1]. Obviously, educational methods, as well as industry technology, have undergone tremendous changes that are somehow reflected in current courses. Basic skills for a petroleum engineer, besides mastering fundamentals of mathematics, physics, and chemistry, will include:

- Geology;
- Well drilling technology;
- Formation evaluation;
- Oil and gas production technology;
- Properties of reservoir rocks;
- Properties of reservoir fluids;
- Fluid flow in porous media;
- Reservoir analysis and management.

❸ There is evidence[2] that, regarding basic technical knowledge, the majority of newly petroleum engineering graduates are well prepared. On the other hand, apparently this is not true with regard to the fast-changing requirements of the oil industry, where there is an expectancy that young professionals will be prepared to exercise leadership, deal with business issues, and implement policies that will contribute to corporate success and profitability. It is probably unrealistic to expect to find the aforementioned set of skills on a recent graduate. However, as a goal, petroleum engineering education should provide the students with the means to use their technical background and personal qualification to acquire those skills after a short period of time

1. Cunha, J. C., & Cunha, L. B. (2005). Looking ahead: Challenges for petroleum engineering education. *Oil and Gas Business*. Retrieved from http://webcache.googleusercontent.com/search?q=cache:pHP536Yb0h4J:ogbus.ru/article/view/looking-ahead-challenges-for-petroleum-engineering-education-3/23308+&cd=1&hl=zh-CN&ct=clnk&gl=jp]

subsequent to graduation.

❹ Clearly, achievement of that goal will depend not only on university infrastructure, laboratory facilities, and well-prepared professors. Fundamental importance should also be given to recruitment of students. Two recent panels[3][4] have reached similar conclusions regarding the fact that attracting the "best and the brightest" students is essential and somewhat is an objective that is not being completely fulfilled by petroleum engineering schools.

Importance of Industry-University Interaction

❺ Over the last twenty years, the field of petroleum engineering has undergone major changes. The evolution of technology, as well as the increasing presence of computerized tools in nearly all stages of the exploration-production process, has generated new needs in the educational system. One frequent comment is that academia not always has evolved fast enough in order to meet those needs.

❻ The perception that universities do not follow, with the necessary fast pace, the trends of the industry (the real world) is present in almost every discipline. Most of the times, this perception does not represent a completely fair view of academia. Universities do not have to concentrate only on the development of mere technical skills. Knowledge of fundamentals of the exact sciences is still very important for engineers and the efforts dedicated to mastering those concepts should not be influenced by new developments and the immediateness of the industry. On the other hand, it must be recognized that academic courses should somehow be influenced and reflect changes undergone by the industry. For that matter, a close collaboration between industry and academia will certainly contribute to better preparing future professionals.

❼ This interaction should go beyond scholarship programs and research funding. Industry leaders interested in improvement of petroleum engineering courses should actively participate in industry-academia seminars clearly pointing out problems detected and aspects in need of enhancement. At the same time, they should also be willing to participate in the efforts for improvement of the courses with suggestions, constructive analysis, talks as guest speakers, and sharing of field data that may be used to provide, for certain courses, a more exciting and industry-related environment[5]. Sharing of actual field data sets is considered one of the most efficient ways[3] for industry to contribute to the enhancement of university education. However, this kind of contribution is relatively rare either due to the absence of mechanisms allowing this action or due to confidentiality matters.

❽ Moreover, a continuous feedback from industry to academia relating how junior professionals are performing and how they are meeting (or not) industry expectations in terms of technical skills and general knowledge of the industry and its business must be provided. Note that this is easier said than done since not always there are appropriate channels allowing free communication between industry leaders and academia. Actually, establishment of such channels

is one of the main challenges to be faced with and should be tackled as a priority for both industry and academia organizations. This is a task that must be embarked upon by academia and industry professionals as well.

Enhancing Preparedness for Professional Life

❾ As mentioned before, knowledge of the fundamentals of exact sciences is of utmost importance for engineers. However, this does not mean that students must have their entire petroleum engineering course devoted to traditional academic disciplines. There are a number of practical measures that may play an important role in preparing students for the challenges that will be faced with after graduation.

❿ Modern courses should include comprehensive laboratory training, involving such fundamental areas as properties of reservoir fluids, fluid flow in pipes and porous media, drilling fluids, rock mechanics, and hydraulics. Theoretical concepts explained in classroom will be better comprehended after practical related applications are performed in laboratory classes. In addition, correlation between theory, laboratory experiments, and actual industry applications will be further enhanced if at the same time an extensive field trip program can be implemented as an auxiliary tool to academic education. Note that this tool, although not mentioned on the previous section, also represents a powerful means of industry-academia interaction. Visits to operation sites, production plants, and industrial laboratories will certainly give students a much better perspective about the industry.

⓫ Current industry trends ask for professionals whose skills supersede traditional academic training. Our industry demands from our possible future leaders and managers a basic understanding of the oil business, the global market, its trends, risks, and economical implications. Knowledge about major oil and service companies and their markets and geographic areas of operation is also necessary since, as it is well known, professional development and promotions often come with re-allocation to a different region or country.

⓬ In addition, modern professionals will be asked to be well informed of legal and ethics issues and have an awareness of matters related to health, safety, and environment. Besides that, even from junior professionals, it will be asked to have good presentation skills and ability to communicate ideas either orally or with written reports in a logical and clear way. A good academic program will have a profusion of assignments related to critical paper reading, report writing, and technical presentations. Also students should be encouraged, whenever possible, to attend conferences, seminars, and technical presentations since examples from well-established professionals are still one of the best incentives to self-improvement.

Challenges on Attracting Students

⓭ As mentioned before, not always we have been successful in attracting high level students to petroleum engineering courses. Besides that, there are also concerns, at least in North America[3], related to the under-representation of women and minorities in the petroleum engineering career.

⓮ One of the reasons often mentioned[4] for a career in the oil business not to be considered attractive enough for young bright minds is related to the somewhat tarnished image of the industry. This image, as unjust as it may be, clearly is present in many segments of our society. Moreover, there is also the fact that exploration and production of oil and gas is frequently seen as an uncharmed, dull activity. We may be failing to publicize the impressive degree of advanced technologies used to find and produce oil and the excitement and challenges present in the career.

⓯ Many universities promote open career days, when students can come to the universities, visit laboratories, and attend presentations. This is a good measure although, apparently, a tradition that is only common in North America.

⓰ Companies and professional societies should also participate in this effort. Information is the key to have more young men and women interested in the career. Besides that, students should be informed about the diversified employment opportunities for petroleum engineers. The number of schools and consequently the number of graduates every year is relatively small when compared with other engineering disciplines, which contributes for constant availability of good job opportunities.

⓱ Another point that is always mentioned by prospective students, and even by petroleum engineering students, is related to the long term subsistence of the industry. Even though there are serious studies[6] indicating that the oil industry will be the major producer of energy for decades to come, the much more publicized studies indicating the end of the oil era for the next few years has obviously created an understandable fear that there is no future for a career in petroleum engineering.

Continuing Education Programs

⓲ This is a tool that is not well utilized by professionals after concluding their BSc programs. Probably one of the reasons for that is the fact that programs offered by petroleum engineering schools are fragmented, not always attending specific professional needs and, most of the times, not efficiently publicized. However, this could be one of the most efficient ways to proficiently promote communication between industry and academia. Often these programs are considered uneconomic but certainly this would not be an issue if there were more participation. On the other hand, obviously, increase in attendance will only occur if such programs become regarded by engineers as true tools for professional development and career advancement.

⓳ Universities, on the other hand, will benefit from the feedback that can be provided by experienced engineers attending the programs. This is a real and tangible benefit and also an indication that analysis of these programs by educational institutions should go beyond the mere

economic point of view.

Conclusions

⑳ Effort should be made towards the goal of attracting more and better prepared students to petroleum engineering schools. The oil industry as well as universities must make every effort in order to better communicate to society the importance of the business for society, its high-tech environment, and its numerous possibilities for a rewarding career.

㉑ Sharing field data and thus allowing more realistic examples and field cases to be presented to students, besides contributing to the increase of the overall quality of petroleum engineering courses, will also better prepare the students for the oil business environment.

㉒ Without compromising traditional science education, regarded as fundamental for engineering students, emphasis should also be placed on a broad range of information related to the oil business and the global economy.

㉓ A well-prepared professional will have to be able to analyze projects not only from a technical standpoint, but also from the perspective of risk and economic analysis. Notions related to ethics, health, safety, and environment are also essential.

㉔ Establishment of channels, allowing free communication between industry and academia, must be a priority in order to achieve efficient and up-to-date courses. Industry feedback is important and certainly petroleum engineering education for the next 20 years will be closely related to the future reserved to the industry.

References

[1] A. E. Uhl, "Petroleum engineering education: The first half-century," *J. Pet. Technol.*, April 1965, pp. 377–386.

[2] H. Kazemi et al., "The fifth SPE colloquium on petroleum engineering education—an industry perspective," paper SPE 64308 presented at the 2000 SPE Annu. Tech. Conf. and Exhibition, Dallas, TX, U.S.A, Oct. 1–4, 2000.

[3] W. J. Lee et al., "Petroleum engineering education: The road ahead," paper SPE 64307 presented at the 2000 SPE Annu. Tech. Conf. and Exhibition, Dallas, TX, U.S.A, Oct. 1–4, 2000.

[4] P. M. Lloyd, and B. F. Ronalds, "Petroleum engineering education and training initiatives across Asia Pacific," paper SPE 84351 presented at the 2003 SPE Annu. Tech. Conf. and Exhibition, Denver, Colorado, U.S.A, Oct. 5–8, 2003.

[5] J. C. Cunha, "Teaching well logging and formation evaluation for petroleum engineering students," *Insite Mag.*, a publication of the Canadian Well Logging Society, June 2004.

[6] World Petroleum Assessment 2000—A Study by the U.S. Geological Survey, http://pubs.usgs.gov/dds/dds-060.

Integrated Exercises

I Read the following academic words, and check whether you can use them appropriately. For those you can not, look up in a dictionary or search online about their contextual use. Write down notes to strengthen your memory.

Reading A

1. bachelor
2. specialty
3. entry-level
4. pursue
5. participate
6. aspiring
7. approve
8. certification
9. faculty
10. competency

Reading B

1. pronounced
2. consultation
3. considerable
4. articulate
5. overall
6. utilization
7. transferable
8. exceptional
9. resolution
10. procedure
11. appetite
12. daunting
13. forefront
14. contractor
15. auditable
16. impediment

Reading C

1. properties
2. undergo
3. expectancy
4. qualification
5. recruitment
6. panel
7. interaction
8. computerize

9. academia
10. detect
11. enhancement
12. constructive
13. confidentiality
14. utmost
15. comprehensive
16. theoretical
17. auxiliary
18. supersede
19. ethics
20. profusion
21. assignment
22. seminar
23. somewhat
24. representation
25. minority
26. tarnish
27. unjust
28. segment
29. publicize
30. subsistence
31. fragment (*vt.*)
32. tangible
33. notion

II Decide on the contextual meaning of the following terms and expressions.

Reading A 

1. hold a bachelor's degree:________________
2. a graduate degree for advanced research positions:________________
3. petroleum engineering careers:________________
4. job outlook:________________
5. analytical skills:________________
6. research laboratories:________________
7. drilling sites:________________
8. postsecondary education:________________
9. civil engineering:________________
10. faculty positions:________________
11. entry-level employment:________________
12. a well-rounded background in general engineering concepts:________________
13. petroleum engineering programs:________________

14. managerial positions:______

Reading B

1. commodity markets:______
2. the frontlines of skill assessment:______
3. dynamic landscape:______
4. be adept with soft skills:______
5. dispute resolution:______
6. hiring managers:______
7. continuing education courses:______
8. introductory and advanced courses:______
9. cloud-based software:______
10. transferring skills:______

Reading C

1. fulfill industry needs:______
2. well drilling:______
3. formation evaluation:______
4. university infrastructure:______
5. actual field data sets:______
6. free communication between industry leaders and academia:______
7. operation sites:______
8. production plants:______
9. professional development and promotions:______
10. legal and ethics issues:______
11. good presentation skills and ability:______
12. critical paper reading:______
13. open career days:______

14. engineering disciplines:______________________________

15. career advancement:______________________________

16. the overall quality of petroleum engineering courses:______________________________

17. traditional science education:______________________________

18. industry feedback:______________________________

III Analyze the grammatical structure of the following complex sentences, figure out the meaning of each sentence, and paraphrase them.

1. Previous exposure to change management principles, taking on new projects that may lie outside your traditional role and enrolling in continuing education courses, are all ways to remain mentally flexible and display an appetite for the dynamic world of oil and gas. (Reading B, Para. 9)

2. Training certifications and compliance are forefront in ensuring required skills are present and in good working order. Regulatory oversight will only continue to grow as energy projects expand in scope, size, and need for public support. (Reading B, Para. 13)

3. The ability to assess obstacles through the lens of imagination, risk-taking and non-traditional assessment models will provide the kind of new, useful, and feasible solutions required. (Reading B, Para. 16)

4. Our Age of Anxiety is, in great part, the result of trying to do today's job with yesterday's tools and yesterday's concepts. (Reading B, Para. 17)

5. On the other hand, apparently this is not true with regard to the fast-changing requirements of the oil industry, where there is an expectancy that young professionals will be prepared to exercise leadership, deal with business issues, and implement policies that will contribute to corporate success and profitability. (Reading C, Para. 3)

6. Knowledge of fundamentals of the exact sciences is still very important for engineers and the efforts dedicated to mastering those concepts should not be influenced by new developments and the immediateness of the industry. (Reading C, Para. 6)

7. At the same time, they should also be willing to participate in the efforts for improvement of the courses with suggestions, constructive analysis, talks as guest speakers, and sharing of field data that may be used to provide, for certain courses, a more exciting and industry-related environment. (Reading C, Para. 7)

8. Note that this is easier said than done since not always there are appropriate channels allowing free communication between industry leaders and academia. Actually, establishment of such channels is one of the main challenges to be faced with and should be tackled as priority for both industry and academia organizations. (Reading C, Para. 8)

Unit 7

Petroleum and Culture

Unit Objectives

Goal 1

Get to know the corporate culture of the oil and gas industry, including its essence, elements, and traits.

Goal 2

Become fully aware of why cultural alignment is of tremendous significance in maintaining the competitive edge in the fierce competition of the global market.

Goal 3

Get informed of how the oil and gas industry keeps culturally cohesive for the healthy and sustainable development through task-based activities.

Besides facing a very complex geopolitical and economic environment, companies in the oil and gas industry face significant challenges related to business ethics and how socially responsible their actions are. These actions are the foundations of corporate culture, a key component of the core competitiveness of a company, and the driving force for its growth. Such corporate culture is conducive to riding over hard times and growing stronger. Oil and gas companies entail developing the corporate culture based on the fine traditions and guided by a vision of innovative, coordinated, green, open, and shared development, so as to achieve sustainable and healthy growth.

Pre-Class Activities

Activity One Search for the Definitions

Search online for the definitions of the following terms or concepts and share your findings with your team members.

1. operational excellence:__________
2. organizational cultures and structures:__________
3. safety culture:__________
4. corporate culture:__________
5. work culture:__________
6. a cohesive work environment:__________
7. business ethics:__________
8. personal and professional development/growth:__________
9. integrated operations:__________
10. corporate social responsibility:__________

Activity Two Watch the Video About the Petroleum Safety Culture

Watch the following video about how to wipe out the risks involved in managing the oil and gas industry and summarize the key points for a team presentation.

 **Video:** Elimenating the Top Five HSE and Quality Risks in Oil and Gas

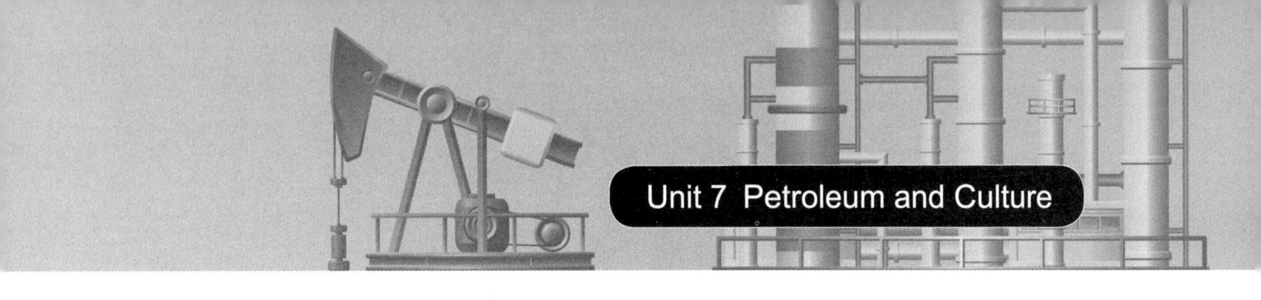

While-Class Activities

Activity One Analyze the Need to Get Culture Right

The following article probes into why there is a cultural mismatch in the oil and gas industry. Summarize the writers' recommendations on how the oil and gas industry should do to fend off the fierce business competition, and make a class presentation.

Reading A

The Cultural Mismatch in the Oil and Gas Industry[1]

Simon Harries and André Baken

❶ When Jack Welch took over GE back in 1981, his top priority was to achieve operational excellence while driving costs down, so he needed a foolproof, cheap, and reliable workforce. What he found were human beings instead of machines; they made mistakes, became sick, asked questions, lost time, and were generally the most unreliable piece in his industrial complex. So he decided to develop a culture that would take human behavior out of the equation as much as possible and bring it as close to machine behavior as he could manage: predictable, repeatable, and submissive to strict quality control. His fear-oriented management style worked well for a long period, other American industries followed his example, and people became a "human resource".

❷ What Jack Welch and his many disciples did not perhaps realize was that cultures go in and out of style, just as technology does. Very soon, a new philosophy of corporate management grew up which would come to undermine the centralized, command and control, hierarchical system put in place at so many large corporations, as they followed the lead set by GE. The dot.com boom of the late 1990s led to the Silicon Valley phenomenon that we see all around us today. Now, the technology field is more or less dominated by corporations that may be large in size but still think and act almost as start-ups. What we see is that the freer, more open, and questioning cultures set by these new age technology companies are much more attractive to disruptive thinkers and innovators than old-fashioned, top-down cultures can possibly be.

1. Harries, S., & Backen, A. (2015, November 30). *The cultural mismatch in the oil and gas industry*. Retrieved from https://www.linkedin.com/pulse/cultural-mismatch-oil-gs-industry-andr%C3%A9-baken

❸ You might ask what does this all mean for the oil and gas industry. The answer is plenty. Oil and gas majors and their service companies adopted or developed their own version of the Welch management vision and practice. They too have command and control cultures, they too tend to see their people as "resources", and they too have a tendency to make them redundant when profits drop, exactly as they would mothball a drilling rig or scrap any piece of redundant machinery. In the long term, companies will pay a price for this approach that might be higher than any short term savings, largely because their lay-off system cannot identify the talent that might have helped them move into the next cycle. They are likely to lose their most creative thinkers, who are very much in demand within ambitious technology companies, and most of them will never come back.

❹ Culture, in other words, is a key factor in determining who will be winners and losers in an increasingly unpredictable market. This is a growing problem across the oil and gas industry. Large service companies identified the need to have top-quality technology innovators within their organizations many years ago, which is when they started to buy and attempt to assimilate specialist software and technology development businesses. We think that this is where a cultural mismatch started to appear, and this has played a big part in destroying value.

❺ When a very large corporation buys a smaller, faster business operating in a field known for its focus on innovation, one of two things will happen:

- Option one: The purchasing company recognizes the need to build on the creative culture of the business they are buying, in which case a kind of "reverse takeover" happens, with the acquired culture permeating and eventually replacing the existing big corporate culture.
- Option two: The acquiring company thinks it is filling a gap in its portfolio and expects the company being bought to adopt the parent culture, work in their niche like good soldiers, and show discipline in following the corporate line.

❻ We think that virtually every technology acquisition in the oil and gas market within the past decade or more has followed the latter model. Fast-moving, creative IT businesses have been expected to fit into the rigid and almost military culture typical of "Jack Welch-style" corporations. Generally speaking, these have not been happy marriages. We say this as a result of simple observation. We can see that traditional oil and gas services companies are not coping well with the huge changes now taking place as a result of the crash in oil prices and the growing political influence driving the actions of large NOCs.

❼ These forces are shaping the market and nowhere is this clearer than in the rise of integrated operations (IO). IO is all about healing the divide between operational technology (OT) and IT, enabling business leaders (and politicians very often) to have a complete view of the truth across

the entire area of operations. To do this, they need to work with creative, confident businesses who are able to think broadly and deliver solid solutions across the entire spectrum of its clients' activities. That means E & P, business systems, analytics, decision support at boardroom level, long-term strategy and do all of these in the context of the macro-economic influences that actually drive the broader industry.

❽ Now, companies that have been acknowledged market leaders for almost a century ought to be perfectly placed to act as the key strategic partners of governments (through NOCs) and global corporates (IOCs and others) in this most sensitive and fast-moving period of our history. But this is not happening. Instead, we see mainstream IT services companies (IBM, HP, even Wipro), global consultancies (Accenture, McKinsey) taking the high ground and making a pitch to be the key strategic partners of choice for a changing industry.

❾ We don't think this is because the traditional players lack capability or insight or know-how or ambition; we think it's because there is a mismatch at the heart of their cultures. In other words, they cannot deliver integrated operations because their own operations and cultures are not integrated. Somehow, the attempt to bring new technology thinking into deeply-rooted, long-established oil and gas companies has not worked. The new cultures have not "taken", serious divisions still exist, and great companies are suddenly looking vulnerable to newcomers like the IT systems integrators (who understand technology transformation in a more profound manner) or feisty specialists (who will always be faster, cheaper, more agile, and almost always, more innovative).

❿ There is time to get this right and to turn round what appears to be a tide that is definitely flowing out. But there's not much time. Our advice? Don't obsess about the technology solutions as yet, get that culture right. That's going to be tougher but far more important.

Activity Two Do Further Reading for Facts or Data

Why is culture a key for the sustainable development and the healthy growth of the oil and gas industry? Reading B will provide the answer. Read it carefully first, hold a seminar discussion with your team members about the reason, and then summarize the top cultural qualities.

Reading B

Why Culture Is Key?[1]

Barry Jaruzelski, John Loehr, and Richard Holman

❶ The elements that make up a truly innovative company are many: a focused innovation strategy, a winning overall business strategy, deep customer insight, great talent, and the right set of capabilities to achieve successful execution. More important than any of the individual elements, however, is the role played by corporate culture—the organization's self-sustaining patterns of behaving, feeling, thinking, and believing—in tying them all together. Yet according to the results of this year's Global Innovation 1000 study, only about half of all companies say their corporate culture robustly supports their innovation strategy. Moreover, about the same proportion say their innovation strategy is inadequately aligned with their overall corporate strategy.

❷ This disconnect, as the saying goes, is both a problem and an opportunity. Our data show that companies with unsupportive cultures and poor strategic alignment significantly underperform their competitors. Moreover, most executives understand what's at stake and what matters, even if their companies don't always seem to get it right. Across the board, for example, respondents identified "superior product performance" and "superior product quality" as their top strategic goals. And they asserted that their two most important cultural attributes were "strong identification with the consumer/customer experience" and a "passion/pride in products".

❸ These assertions were confirmed by innovation executives we interviewed for the study. Fred Palensky, executive vice president of research and development and chief technology officer (CTO) at innovation leader 3M Company, for example, puts it this way: "Our goal is to include the voice of the customer at the basic research level and throughout the product development cycle, to enable our technical people to actually see how their technologies work in various market conditions."

❹ If more companies could gain traction in closing both the strategic alignment and culture gaps to better realize these goals and attributes, not only would their financial performance improve, but the data suggest that the potential gains might be large enough to improve the overall growth rate of the global economy.

❺ To that end, we continue to emphasize the key finding that our Global Innovation 1000

1. Jaruzelski, B., Loehr, J., & Holman, R. (2011, October 25). *Why culture is key?* Retrieved from https://www.strategyand.pwc.com/media/file/Strategyand-Global-Innovation-1000-2011-Culture-Key.pdf

study of the world's biggest spenders on research and development has re-affirmed in each of the past seven years: There is no statistically significant relationship between financial performance and innovation spending, in terms of either total R & D dollars or R & D as a percentage of revenues. Many companies—notably, Apple—consistently underspend their peers on R & D investments while outperforming them on a broad range of measures of corporate success, such as revenue growth, profit growth, margins, and total shareholder return. Meanwhile, entire industries, such as pharmaceuticals, continue to devote relatively large shares of their resources to innovation, yet end up with much less to show for it than they—and their shareholders—might hope for.

❻ Last year, we looked at the innovation capability sets companies put together, how they vary by innovation strategy, and which groups of capabilities can best enable companies to outperform their peers. This year, we took a different vantage point, analyzing the ways that critical organizational systems and cultural attributes support those capability sets that are most likely to promote innovation success. The results suggest that the ways R & D managers and corporate decision makers think about their new products and services—and how they feel about intangibles such as risk, creativity, openness, and collaboration—are critical for success. As part of this year's study, we surveyed almost 600 innovation leaders in companies around the world, large and small, in every major industry sector. As noted, almost half of the companies reported inadequate strategic alignment and poor cultural support for their innovation strategies. Possibly even more surprising, nearly 20% of companies said they didn't have a well-defined innovation strategy at all.

❼ Understanding these issues is particularly important now that innovation spending is on the rise again. After last year's 3.5% drop in global innovation spending, the first-ever decline shown in the data we have tracked for more than a decade, R & D outlays have recovered. Spending among the Global Innovation 1000 surged 9.3% in 2010, thanks in great part to the perception of a worldwide economic recovery.

The Alignment Gap

❽ Issues of culture have long been of great concern to corporate executives and management theorists alike, whether they apply to companies as a whole or to selected areas such as innovation. The reason is obvious: culture matters, enormously. Studies have shown again and again that there may be no more critical source of business success or failure than a company's culture—it trumps strategy and leadership. That isn't to say that strategy doesn't matter, but rather that the particular strategy a company employs will succeed only if it is supported by the appropriate cultural attributes. So when we approached the topic of culture in the context of innovation for this year's study, our primary goals were to determine which cultural attributes were most critical to underpinning the focused capability sets required for each distinct innovation strategy that we have previously identified.

❾ The results are clear—and may explain why many companies have difficulty making their substantial R & D investments pay off. Overall, 36% of all respondents to our survey admitted that their innovation strategy is not well aligned to their company's overall strategy, and 47% said their company's culture does not support their innovation strategy. Not surprisingly, companies saddled with both poor alignment and poor cultural support perform at a much lower level than well-aligned companies. In fact, companies with both highly aligned cultures and highly aligned innovation strategies have 30% higher enterprise value growth and 17% higher profit growth than companies with low degrees of alignment. (See Exhibit 1)

Exhibit 1: The Alignment Advantage

Only 44% of companies surveyed have both highly aligned cultures and highly aligned innovation strategies, and it pays off in performance: they outperform on growth in both profits and enterprise value.

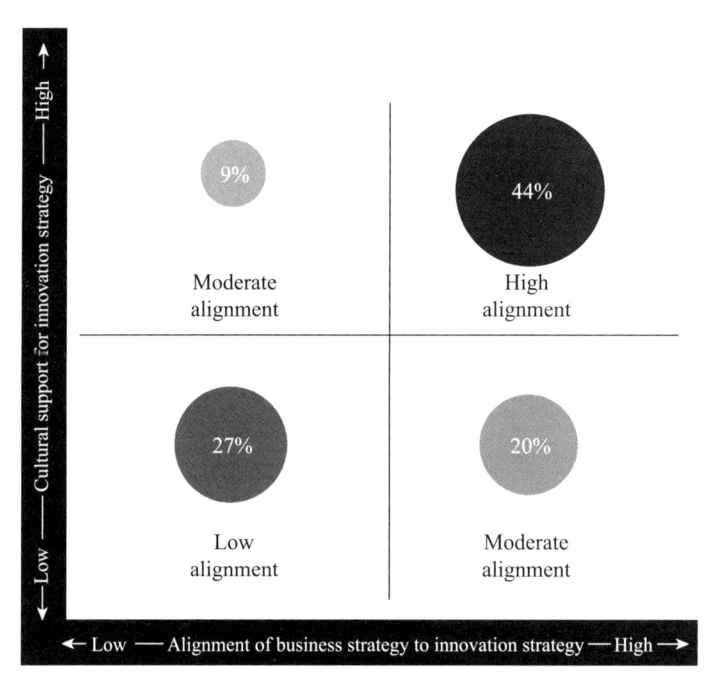

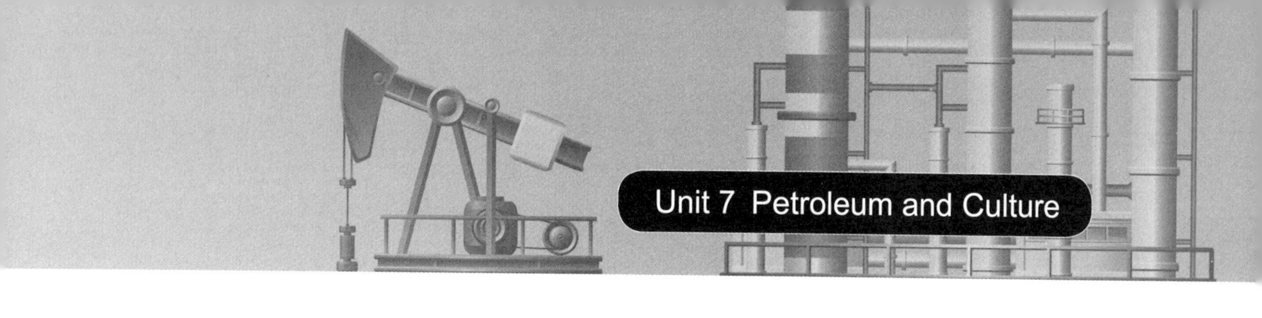

(Continued)

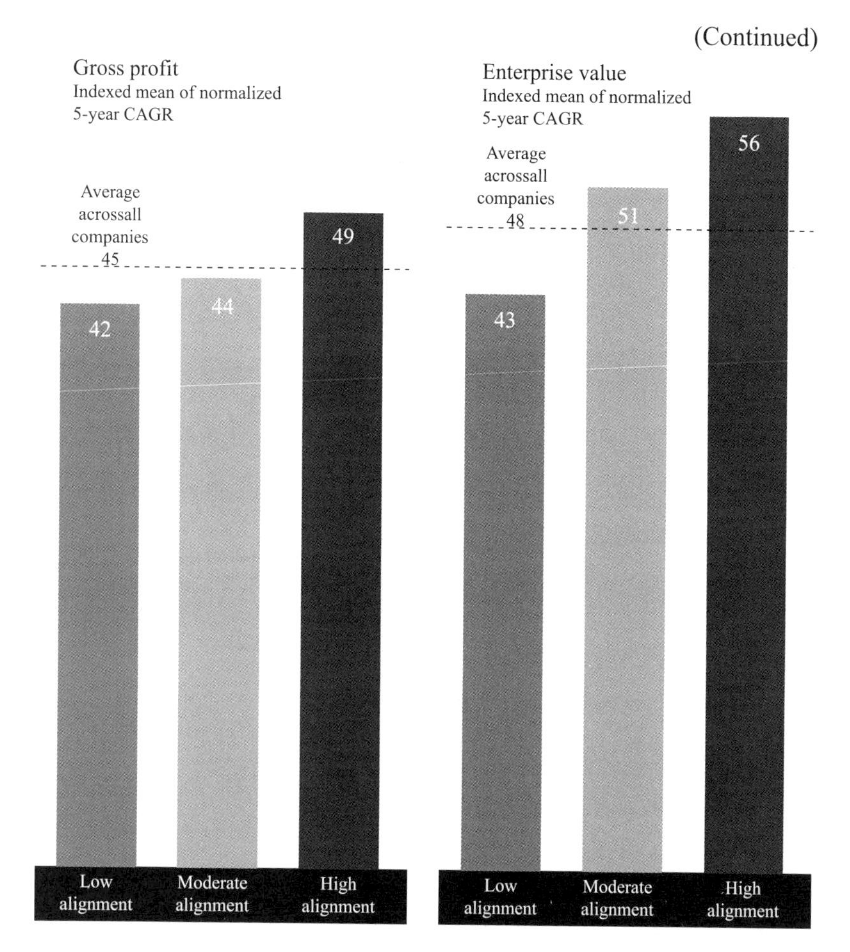

Figure 7-1 Innovation strategy and cultural alignment matrix

(Source: Booz & Company.)

⑩ On the other hand, companies whose strategic goals are clear, and whose cultures strongly support those goals, possess a huge advantage. 3M is a case in point. Palensky articulates his company's innovation strategy clearly: "We call it 'customer-inspired innovation'. Connect with the customer, find out their articulated and unarticulated needs, and then determine the capability at 3M that can be developed across the company that could solve that customer's problem in a unique, proprietary, and sustainable way."

⑪ Culture plays a critical role in this strategy, says Palensky: "For over 100 years, 3M has had a culture of interdependence, collaboration, even codependence. Our businesses are all interdependent and collaboratively connected to each other, across geographies, across businesses, and across industries. The key is culture."

⑫ Despite their differences in performance, most companies strongly agree on the strategic goals that matter most in achieving innovation success: "Superior product performance" and "superior product quality" were ranked number one or two by a plurality of more than 40% of all respondents. Other goals, such as "developing low-cost products" and "speed-to-market", were given much lower priority. (See Exhibit 2)

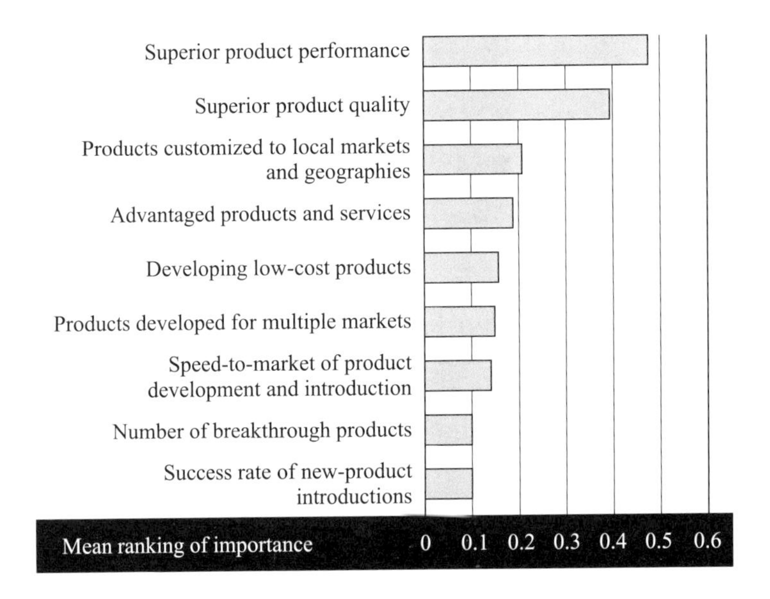

Figure 7-2 Perceived importance of common innovation goals/outcomes

(Source: Booz & Company.)

⑬ Similarly, companies agree strongly on the cultural attributes that are most prevalent at their companies. More than 60% cited "strong identification with the customer" as among the top two, and 50% chose "passion for and pride in products". The lowest ranked was "tolerance for failure in the innovation process". (See Exhibit 3) This finding, which contradicts some of the academic research on the subject, raises serious questions about companies' real appetite for risk taking in their innovation practices.

Exhibit 3: Top Cultural Attributes

Companies agreed strongly on the two most important cultural attributes. There was less unanimity on the importance of other attributes with very few citing tolerance for failure in innovation as essential.

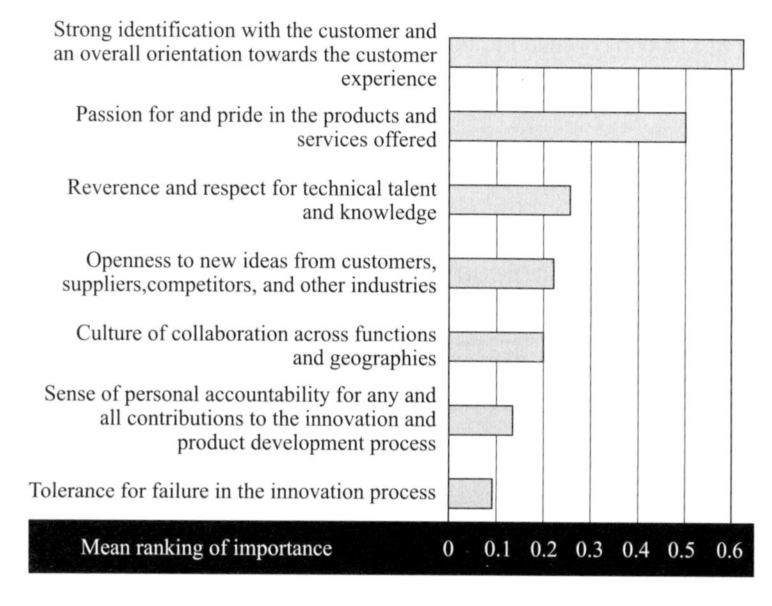

Figure 7-3 Perceived importance of cultural attributes
(Source: Booz & Company.)

⑭ In general, companies also continue to show a range of significant gaps in how their strategic goals and cultural attributes contribute to performance and support their innovation. Companies that underperform their peers have much to gain if they can close these gaps and achieve much higher degrees of cultural and strategic alignment. We believe the way to do so lies in gaining a greater understanding of the cultural attributes that any given company needs to foster, given its particular innovation strategy. Soma Somasundaram, executive vice president of the Fluid Management segment at the Dover Corporation, describes the challenge this way: "Poor innovation performance is usually not caused by a lack of ideas or lack of aspirations. What some companies lack is the structure needed to effectively dedicate resources to innovation. It's the lack of will to develop a strategy that can balance today's need versus tomorrow's."

After-Class Activities

Activity One Watch the Video About Business Ethics

Business ethics apply to all aspects of business conduct and are relevant to the conduct of individuals and organizations. Watch the following video, take down notes on how the oil giants put emphasis on their ethical behavior, and share your notes with your team members.

 **Video:** Texas Oil Empire: KXAN Drills into Ethics of Industry Regulators

Activity Two Approach Petroleum Organizational Culture

Today's globalized nature of competitiveness in oil and gas markets is placing more pressure on organizations operating in these industries to develop more effective organizational cultures and structures, which can be regarded as an integral part of petroleum culture. Read the article below and summarize the key organizational elements that affect the performance of oil and gas companies.

Reading C

How Organizational Elements Influence Performance in Oil and Gas Companies[1]

Mostafa Sayyadi

❶ Today's globalized nature of competitiveness in oil and gas markets is placing more pressure on companies operating in these industries to develop more effective organizational culture, structure, strategy, and information technology. There are many studies that focus on the organizational and managerial factors that drive organizational performance.

❷ Organizational culture, structure, strategy, and information technology are such areas that play a critical role and are strategic prerequisites for business success in today's hypercompetitive

1. Sayyadi, M. (2017, February 07). *How organizational elements influence performance in oil and gas companies*. Retrieved from http://www.consultingmag.com/sites/cmag/2017/01/13/how-organizational-elements-influence-performance-in-oil-and-gas-companies/?slreturn=20190717110045

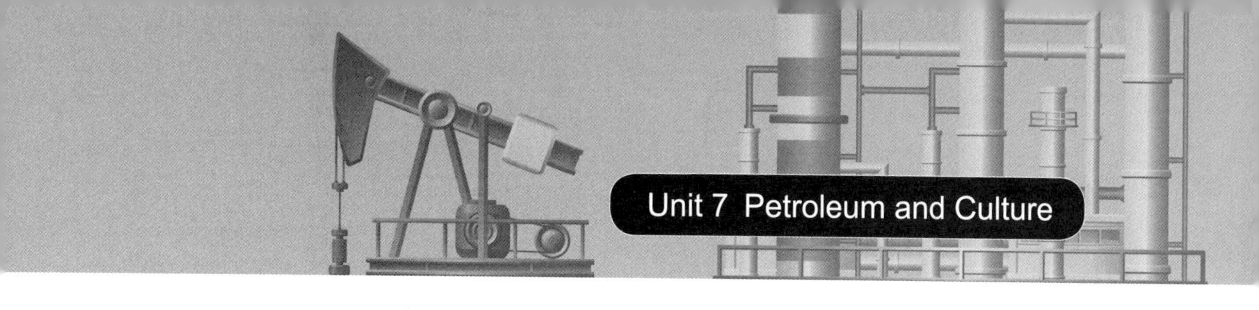

environment of oil and gas markets. I place a new emphasis on these organizational factors, and shed light on these important organizational elements to build effective companies operating in oil and gas industries.

❸ Andrew Pettigrew initially introduced the term "organizational culture" into the business literature in 1979's "On Studying Organizational Cultures". Edgar Schein describes organizational culture as a pattern of shared basic assumptions that the group learned as it solved its problems of external adaptation and internal integration that has worked well enough to be considered valid and, therefore, to be taught to new members as the correct way to perceive, think, and feel in relation to those problems in 1984's "Coming to a New Awareness of Organizational Culture".

❹ Organizational culture is, therefore, reflected in shared assumptions, symbols, beliefs, values, and norms that specify how employees understand problems and appropriately react to them. To analyze the relationship between corporate culture and firm performance, organizational culture could be visualized by its three major aspects, including collaboration, trust, and learning.

❺ Both cultural aspects of collaboration and trust positively contribute to oil and gas companies to effectively and actively respond to environmental changes, customer needs, and employee growth needs through developing effective learning workplaces within these companies.

❻ This also helps these companies to improve performance in terms of the quality of products and services. Learning culture as another cultural aspect sheds light on organizational capabilities to develop learning. It is quite understandable that this cultural aspect can particularly facilitate performance in oil and gas companies, by developing suitable workplaces for experts and internal consultants to effectively share their knowledge with others.

❼ People in fact recognize how old resources can address new and problematic situations by sharing their knowledge within these companies, and this can help to create more innovative ideas for organizational problems. David Maister in his book, *Managing the Professional Service Firm*, says that innovative ideas generation can improve profitability for companies. Thus, I suggest that business consultants in oil and gas industries should consider organizational culture as an important enabler to enhance financial and non-financial performances.

❽ Less emphasis on centralized structures develops communications within oil and gas companies. This less emphasis on centralization also creates more appropriate and effective workplaces for developing learning and growth that in turn can improve sustainable competitive advantage for these companies. This effective learning environment is a by-product of the delegation of authorities that in turn can inspire people to actively participate in organizational decisions.

❾ Decentralization within oil and gas companies can also enable these companies to identify changes in external environment and then help them to actively and effectively respond to these rapid changes. Less emphasis on formalization can also provide freedom for experts and internal consultants to more innovatively handle their work operations, which leads to higher job satisfaction within oil and gas companies.

⑩ James Hesket and his colleagues in their book, *The Ownership Quotient: Putting the Service-Profit Chain to Work for Unbeatable Competitive Advantage*, state that job satisfaction can stimulate the quality of products and services that potentially leads to higher degrees of customer satisfaction and profitability. Therefore, I suggest that business consultants in oil and gas industries should also consider the importance of organizational structure in improving performance in oil and gas companies.

⑪ Organizational strategy can also play a critical role in improving performance. Firms' strategy can be categorized into two prominent streams: The first stream is that research indicates that there is a strong alignment between business strategy and external environment that potentially leads to higher degrees of performance at the organizational level; the other stream sheds light on different typologies of business strategies, and argues that one typology of these existing typologies can create better results for organizations when compared to others. It can be seen that these two streams have highlighted business strategy as an important enabler to improve organizational performance.

⑫ In terms of STROBE strategy dimensions, analysis strategy can in turn develop opportunities for human resources development within organizations operating in oil and gas industries, by assessing current situations in details. Analytical orientation has major effects on the performance of these companies through focusing on analytical decision-making process. Human resources development can potentially facilitate financial performance of oil and gas companies, by improving profitability for these companies.

⑬ Defensive strategy as a necessary requisite enhances profitability, which enhances efficiency in companies' current positions in the hypercompetitive oil and gas markets. Since pro-activeness manifests itself in behaviors such as continuously exploring the emerging opportunities to invest, this strategy can positively contribute to the efficiency of oil and gas companies through helping these companies to find better opportunities for investment that potentially leads to better financial performance for these companies in terms of return on investment (ROI) and profitability.

⑭ Futurity strategy, which implements basic studies to develop an effective and comprehensive vision for the future, can also enable oil and gas companies to identify and actively respond to the changes occurred in the external environment. In line with this, I suggest business consultants in oil and gas industries should consider the critical role of strategy in improving organizational performance for companies operating in these industries.

⑮ Information technology is a key factor to improve business. *Forbes*' reports on American industries clearly indicate that effective information technology significantly contributes to firms' financial performance. These researches acknowledge that information technology is an important

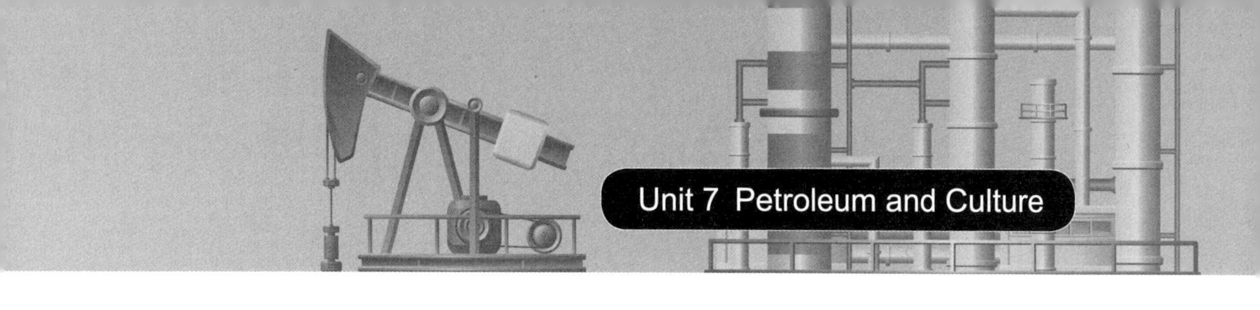

enabler to effectively implement organizational processes. Communication technologies can in fact reduce paper-based transactions for oil and gas companies that potentially decrease costs and subsequently improve profitability within these companies.

⑯ Furthermore, it can be seen that communication technologies contribute to these companies to effectively identify opportunities in the external business environment that leads to identify best opportunities for investment in oil and gas industries that potentially leads to improve financial performance for companies operating in these industry in terms of return on investment.

⑰ Decision-aid technologies as another kind of information technology can also help oil and gas companies to effectively create more innovative solutions for their organizational problems. In this way, I argue that information technology is positively associated with two important factors of product and service quality and customer satisfaction. I, therefore, recommend that oil and gas companies should consider information technology as a key player in improving their performance in today's oil and gas hypercompetitive environment.

Activity Three Practice Summary Writing

Safety culture is also an integral part of petroleum culture. Reading D will present five ways to enhance safety in the oil and gas industry. Read it carefully and write a summary with not more than 150 words.

Reading D

Five Ways the Oil and Gas Industry Promotes a Strong Safety Culture[1]

Carrie Jordan

President, DJ Basin Safety Council

❶ I work as a safety consultant in many industries, including oil and gas, construction, manufacturing, trucking, and underground utilities. Working with many different industries, I get to engage in different safety cultures. I see a widespread willingness to ensure safety, provide training and equipment in the oil and gas industry.

1. Jordan, C. (unknown publication date). *Five ways the oil and gas industry promotes a strong safety culture.* Retrieved from http://www.coga.org/five-ways-oil-gas-industry-promotes-strong-safety-culture/

❷ There are many ways to ensure safety is at the forefront of oil and gas operations every day, including state and federal regulations designed to protect employees and the environment. In my experience, these are five ways that assist companies working on a site in fostering a strong safety culture.

Employee Orientation and Personal Protective Equipment

❸ Mandatory orientations for each employee are required prior to employees even being allowed to enter the location, such as the SafeLand U.S.A Basic Orientation, and operator specific orientations for each company. All sites require mandatory personal protective equipment, including safety glasses, hard hats, steel toed boots, and fire resistant clothing—if you do not wear the right stuff, you are not allowed to the site. During the downturn in oil and gas, I have witnessed employees moving into construction companies and taking these small but meaningful safety habits with them—a testament to this industry's strong safety culture.

Job Safety Analysis

❹ Most companies perform daily task reviews called job safety analysis forms (JSA's) that must be completed prior to any work being conducted. JSA's have to be reviewed by all employees to ensure everyone is on the same page on job steps, hazards, and how those hazards are controlled. This ten-minute exercise allows for communication, questions, and ensures everyone is working together.

Issuing Stop Work Authority

❺ Stop Work Authority allows anyone who sees a potential safety hazard to bring it to the attention of everyone on site and stop work to correct the issue before work resumes. While not always popular to stop the job, issuing a stop work order can protect someone from a hazard or save a life.

Reviewing Near Misses and Injuries

❻ Unfortunately, injuries or near miss incidents do occur. When these do happen, employers, producers, and subcontractors share that information with their networks and pass it on to organizations like the DJ Basin Safety Council to distribute and talk about so we can learn from others. This proactive process of review helps safeguard others who may encounter similar situations.

Vetting Subcontractors

❼ Safety is at the forefront of nearly every operation, every day, for companies working on an oil and gas site. Subcontractors are selected through a safety record evaluation process to ensure companies that gain contracts to work on site have a safety program in place. Contractor management and oversight is taken very seriously. If you do not have a good safety record, you are not going to be selected to work.

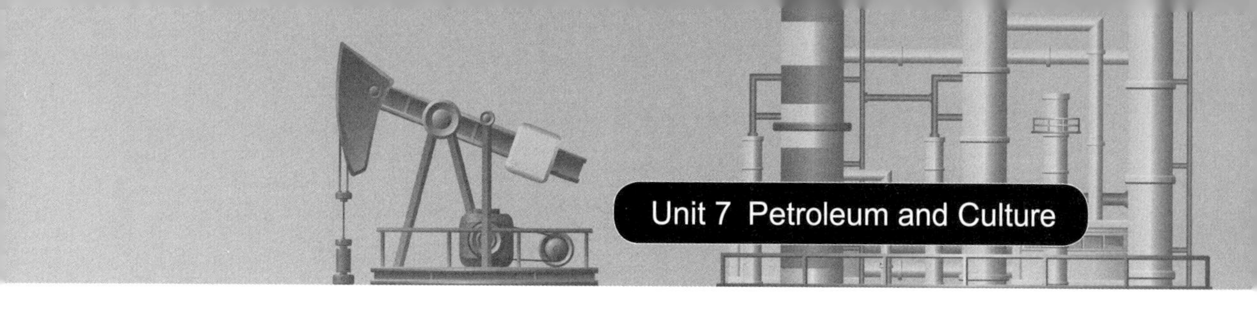

❽ All industries and work types have some type of risk involved and not all companies or industries take safety seriously or make it an inherent part of every day's operations. That's not the case in the oil and gas industry. I have seen an industry wide acceptance and willingness to ensure the overall safety culture in one which I am proud to be a part of.

Integrated Exercises

I Read the following academic words, and check whether you can use them appropriately. For those you can not, look up in a dictionary or search online about their contextual use. Write down notes to strengthen your memory

Reading A

1. foolproof
2. unreliable
3. equation
4. submissive
5. orient
6. hierarchical
7. disruptive
8. centralize
9. vision
10. tendency
11. redundant
12. largely
13. scrap
14. ambitious
15. assimilate
16. option
17. permeate
18. portfolio
19. rigid
20. crash
21. heal
22. spectrum
23. analytic
24. acknowledge
25. mainstream
26. consultancy
27. know-how
28. profound
29. agile
30. obsess

Reading B

1. execution
2. self-sustaining
3. robustly
4. underperform
5. assertion
6. executive

7. traction
8. reaffirm
9. notably
10. peer
11. pharmaceutical
12. outlay
13. trump
14. underpin
15. alignment
16. strategic
17. interdependence
18. codependence
19. proprietary
20. plurality
21. prevalent
22. contradict
23. foster
24. aspiration
25. versus

Reading C

1. competitiveness
2. prerequisite
3. shed
4. valid
5. external
6. specify
7. appropriately
8. visualize
9. consultant
10. problematic
11. enabler
12. by-product
13. delegation
14. inspire
15. decentralization
16. formalization
17. quotient
18. unbeatable
19. prominent
20. typology
21. highlight
22. dimension
23. requisite
24. transaction
25. manifest
26. implement

Reading D

1. mandatory
2. orientation
3. stuff
4. testament
5. hazard
6. subcontractor
7. oversight
8. inherent
9. proactive
10. acceptance

II Decide on the contextual meaning of the following terms and expressions.

Reading A

1. cultural mismatch:________________

2. quality control:________________

3. fear-oriented management style:________________

4. human resources:________________

5. corporate management:________________

6. top-down cultures:________________

7. organizational performance:________________

8. defensive strategy:________________

9. lay-off system:________________

10. top-quality technology innovator:________________

11. specialist software and technology development businesses:________________

12. creative culture:________________

13. corporate culture:________________

14. technology acquisition in the oil and gas market:________________

15. decision support at boardroom level:________________

16. macro-economic influence:________________

17. strategic partner:________________

18. the rise of integrated operations:________________

19. mainstream IT services:________________

20. technology transformation:________________

Reading B

1. a focused innovation strategy:________________

2. business strategy:________________

3. corporate strategy:____________________

4. strategic alignment:____________________

5. customer insight:____________________

6. cultural attributes:____________________

7. strong identification with the consumer:____________________

8. product development cycle:____________________

9. the overall growth rate of the global economy:____________________

10. R & D investments:____________________

11. revenue growth:____________________

12. total shareholder return:____________________

13. the innovation capability sets:____________________

14. vantage point:____________________

15. a worldwide economic recovery:____________________

16. major industry sector:____________________

17. global innovation spending:____________________

18. the focused capability sets:____________________

19. highly aligned cultures:____________________

20. highly aligned innovation strategies:____________________

21. customer-inspired innovation:____________________

22. superior product performance:____________________

23. superior product quality:____________________

24. tolerance for failure in the innovation process:____________________

Reading C

1. managerial factors:____________________

2. organizational culture:____________________

3. external adaptation:____________________

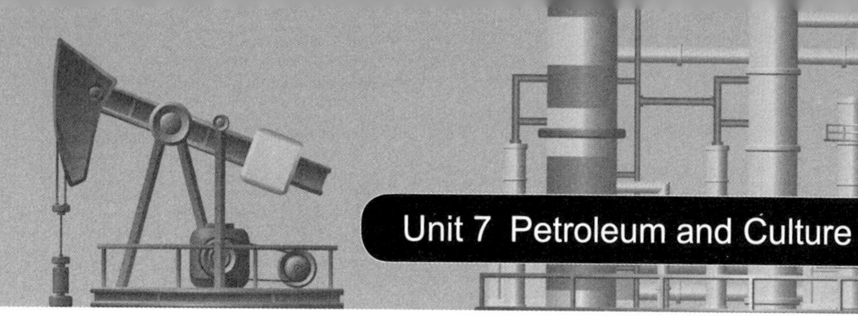

4. internal integration:________________________________

5. employee growth needs:________________________________

6. work operations:________________________________

7. internal consultants:________________________________

8. innovative ideas generation:________________________________

9. the delegation of authorities:________________________________

10. external environments:________________________________

11. job satisfaction:________________________________

12. competitive advantage:________________________________

13. customer satisfaction:________________________________

14. organizational structure:________________________________

15. analytical decision-making process:________________________________

16. financial performance:________________________________

17. return on investment:________________________________

18. analytical orientation:________________________________

19. information technology:________________________________

20. futurity strategy:________________________________

21. paper-based transaction:________________________________

22. decision-aid technology:________________________________

23. service quality:________________________________

Reading D

1. underground utilities:________________________________

2. safety culture:________________________________

3. employee orientation:________________________________

4. operator specific orientations:________________________________

5. hard hats:________________________________

6. steel toed boots:______________________________

7. construction company:______________________________

8. safety analysis forms:______________________________

9. potential safety hazard:______________________________

10. near miss incident:______________________________

11. a safety record evaluation process:______________________________

II Analyze the grammatical structure of the following complex sentences, figure out the meaning of each sentence, and paraphrase them.

1. Now, the technology field is more or less dominated by corporations that may be large in size but still think and act almost as start-ups. What we see is that the freer, more open, and questioning cultures set by these new age technology companies are much more attractive to disruptive thinkers and innovators than old-fashioned, top-down cultures can possibly be. (Reading A, Para. 2)

2. Large service companies identified the need to have top-quality technology innovators within their organizations many years ago, which is when they started to buy and attempt to assimilate specialist software and technology development businesses. We think this is where a cultural mismatch started to appear, and this has played a big part in destroying value. (Reading A, Para. 4)

3. Option one: The purchasing company recognizes the need to build on the creative culture of the business they are buying, in which case a kind of "reverse takeover" happens, with the acquired culture permeating and eventually replacing the existing big corporate culture. (Reading A, Para. 5)

4. The new cultures have not "taken", serious divisions still exist, and great companies are suddenly looking vulnerable to newcomers like the IT systems integrators (who understand technology transformation in a more profound manner) or feisty specialists (who will always be faster, cheaper, more agile, and almost always, more innovative). (Reading A, Para. 9)

5. If more companies could gain traction in closing both the strategic alignment and culture gaps to

better realize these goals and attributes, not only would their financial performance improve, but the data suggests that the potential gains might be large enough to improve the overall growth rate of the global economy. (Reading B, Para. 4)

6. Meanwhile, entire industries, such as pharmaceuticals, continue to devote relatively large shares of their resources to innovation, yet end up with much less to show for it than they—and their shareholders—might hope for. (Reading B, Para. 5)

7. So when we approached the topic of culture in the context of innovation for this year's study, our primary goals were to determine which cultural attributes were most critical to underpinning the focused capability sets required for each distinct innovation strategy that we have previously identified. (Reading B, Para. 8)

8. This effective learning environment is a by-product of the delegation of authorities that in turn can inspire people to actively participate in organizational decisions. (Reading C, Para. 8)

9. Stop Work Authority allows anyone who sees a potential safety hazard to bring it to the attention of everyone on site and stop work to correct the issue before work resumes. (Reading D, Para. 5)

Unit 8

Petroleum and Renewable Energy

Unit Objectives

Goal 1

Get familiar with the current energy portfolio and the energy tilt from fossil fuels to renewable energy.

Goal 2

Get to know how the oil majors balance fossil fuels and renewable energy, the risk and the return.

Goal 3

Get informed of the future energy trends through task-based activities.

Renewable energy is collected from renewable resources, which are naturally replenished on a human timescale, such as sunlight, wind, rain, tides, waves, and geothermal heat. Renewable energy often provides energy in four important areas: electricity generation, air and water heating/cooling, transportation, and rural (off-grid) energy services. Based on REN21's (Renewable Energy Policy Network for the 21st Century) 2017 report, renewables contributed 19.3% to humans' global energy consumption and 24.5% to their generation of electricity in 2015 and 2016, respectively. Worldwide investments in renewable technologies amounted to more than $286 billion in 2015, with countries such as China and the United States heavily investing in wind, hydro, solar, and biofuels. Although the oil majors have no plans to ditch their traditional business, corporate sustainability pressures and the attractiveness of renewables are pushing many towards clean energy. Shell, Total, Statoil, even Exxon—they're all at it. But are the recent moves into solar and wind power lip service, fashion, or a real shift away from fossil fuels?

Pre-Class Activities

Activity One Search for the Definitions

Search online for the definitions of the following terms or concepts and share your findings with your team members.

1. renewable and non-renewable energy:____________________
2. low carbon lifestyle:____________________
3. global energy consumption:____________________
4. energy mix:____________________
5. carbon footprint:____________________
6. the fossil fuel dilemma:____________________
7. the symbiotic relationship of fossil fuels and renewables:____________________
8. sustainable development:____________________
9. Equator Principles:____________________

Activity Two Watch the Video About Energy Sources

Watch the following video about two kinds of energy sources: fossil fuels vs renewable energy, summarize the key characteristics of the two kinds, and share with your team members.

 **Video:** Fossil Fuels vs Renewable Energy

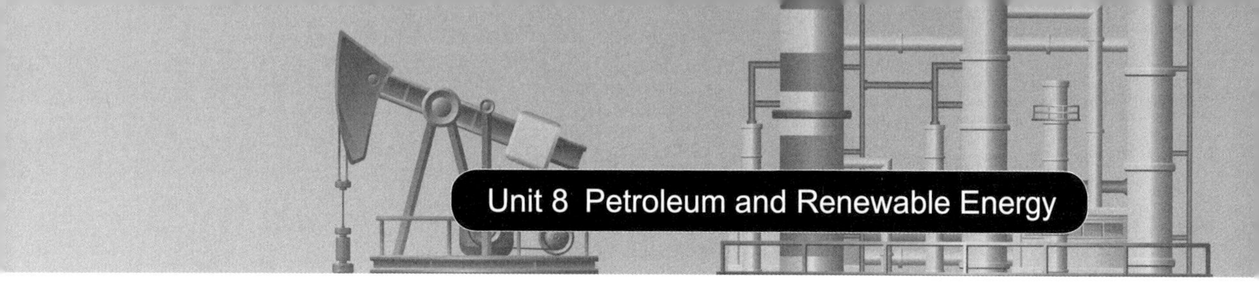

Activity Three Search for Scientific References About Energy Transition

Literature study has suggested that there are momentous changes under way in the global energy system, shifting from oil, gas, and coal towards renewables. Search for some references with regard to the following questions: Why is there such a call for this shift? What role do fossil fuels play in such a big transition? When can we possibly anticipate looking to near 100% renewable scenarios? Summarize your literature study you've done regarding these questions and prepare for a team presentation.

While-Class Activities

Activity One Comment on Literature Study About Energy Transition

You are asked to make a team presentation about your literature study first. Then refer to Reading A to check how much your literature study can coincide with the writer's arguments.

Reading A

Balance of Power Tilts from Fossil Fuels to Renewable Energy[1]

E. Crooks

❶ These are strange days in the energy business. Startling headlines are emerging from the sector that would have seemed impossible just a few years ago.

❷ The Dubai Electricity and Water Authority said in May it had received bids to develop solar power projects that would deliver electricity costing less than three cents per kilowatt hour. This established a new worldwide low for the contracted cost of delivering solar power to the grid—and is priced well below the benchmark of what the Emirates and other countries typically pay for electricity from coal-fired stations.

❸ In the U.K., renowned for its miserable overcast weather, solar panels contributed more power to the grid than coal plants for the month of May.

❹ In energy-hungry Los Angeles, the electricity company AES is installing the world's largest

1. Crooks, E. (2016, July 26). *Balance of power tilts from fossil fuels to renewable energy*. Retrieved from https://www.ft.com/content/8c28d1c2-2400-11e6-9d4d-c11776a5124d

battery, with capacity to power hundreds of thousands of homes at times of high demand, replacing gas-fired plants which are often used at short notice to increase supply to the grid.

❺ Trina Solar, the Chinese company that is the world's largest solar panel manufacturer, said it had started selling in 20 new markets last year, from Poland to Mauritius and Nepal to Uruguay.

❻ It is not only renewable energy that is throwing out such remarkable news. Production costs in the U.S. shale oilfields have been cut by up to 40% in the past two years, according to Wood Mackenzie, the research company. Cargoes of liquefied natural gas have been heading from the U.S. to the Gulf, making the surplus in North America available to the markets of Dubai and Kuwait even though they sit within the world's largest oil and gas producing region.

❼ The implication of those stories is to suggest there are momentous changes under way in the global energy system, undermining received wisdom in the sector. It is clear that the world is shifting towards renewables and—as a proportion of total consumption—away from oil, gas, and coal.

❽ Within the markets for fossil fuels, some sources such as gas are becoming favored over others such as coal. The question for policymakers and industry experts is how far and how fast these changes can go.

❾ Down the decades, an attitude of cynicism in the face of the latest trends has generally been the smart position to take on energy. Assets such as oilfields and power plants are big investments that have operational lives lasting for many decades, and so the fuel mix and fleet of power-generating assets turn over slowly.

❿ Spencer Dale, chief economist at BP, published a fascinating chart in June showing the rate of adoption of existing energy sources and technologies, which makes clear that it is often a lengthy process. For example, in 1899 gas provided just 1% of the world's primary energy needs. Five decades later, that figure had grown to 8%.

⓫ While renewable energy has been growing fast, it is coming from a very low base. "Modern renewables"—mostly biofuels, wind, and solar, but not hydro or traditional biomass—provided just 2.5% of the world's primary energy last year, according to BP.

⓬ That said, there are examples from history of when energy systems have changed rapidly after reaching tipping points. Oil consumption had been growing steadily through the late 19th and early 20th centuries, but really took off during and after the First World War, as warships switched from coal to fuel oil and armies became mechanized with petrol- and diesel-engine vehicles.

⓭ Nuclear power had a similar surge between the Arab oil embargo against the U.S. and other countries in 1973 and the Chernobyl accident of 1986.

⓮ Government policies to address the threat of climate change are today's equivalent.

⓯ The commitments to take action to combat climate change made by 195 countries at the

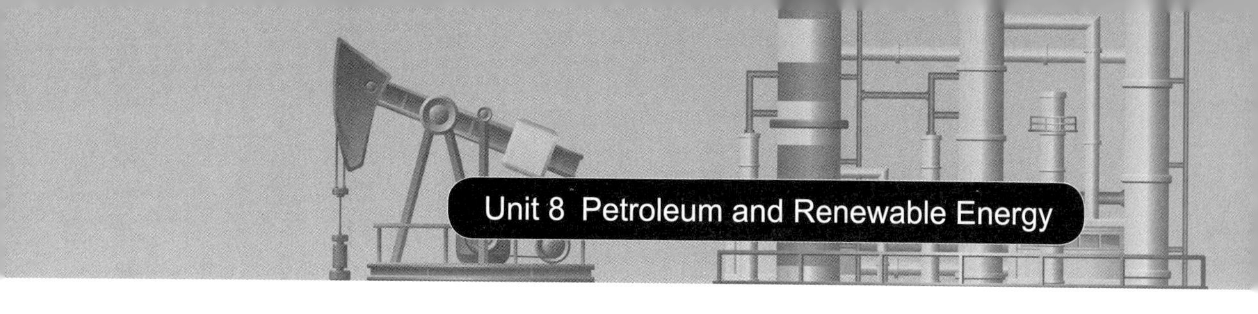

Paris talks at the end of last year are a sign that, however contentious the issue may be politically in the U.S., on a global scale the pressure is unlikely to dissipate any time soon.

⑯ This special report includes examples of innovative technologies that could bring further change to parts of the energy industry. Small modular nuclear reactors, for example, intended to avoid the staggering cost of their larger rivals, are being proposed for use in the U.S. or the U.K. by 2025.

⑰ At the same time, fossil fuel companies are making strides in their efforts to remain competitive. This is not easy. Not only have oil and gas prices plunged over the past two years, but in the long term weaker demand and more abundant supply are expected. Valuations of companies in this sector have been badly dented.

⑱ Some new energy technologies, meanwhile, are not making much progress, such as the development of power plants that capture and store the carbon dioxide they produce. It is commonly assumed among policymakers that carbon capture has become essential if humankind is to enjoy the benefits of fossil fuels while avoiding their polluting effects.

⑲ It is clear, too, that the growth of renewables and other low-carbon energy sources will not follow a straight line. Investment in clean energy has been faltering this year after hitting a record in 2015, according to Bloomberg New Energy Finance. For the first half of 2016, it is down 23% from the equivalent period last year.

⑳ Even so, the elements are being put in place for what could be a quite sudden and far-reaching energy transition, which could be triggered by an unexpected and sustained surge in oil prices. If China or India were to make large-scale policy commitments to electric vehicles, they would have a dramatic impact on the outlook for oil demand.

㉑ In Ernest Hemingway's *The Sun Also Rises*, a character says when he went bankrupt: "Two ways. Gradually, then suddenly." There is a chance that a profound shift in our energy system could sneak up on us in the same way.

Activity Two Balance Renewables and Petroleum Energy

Reading B below describes how the oil and gas sector gets along when the momentum for clean energy is mounting. Read it carefully and summarize the key processes where renewables have gained acceptance and the key measures oil and gas companies can take with due diligence to mediate between renewables and petroleum. Write down your summary and present it to the whole class.

Reading B

When Renewables Meet the Oil and Gas Industry, Opposites Attract[1]

Jason Switzer

❶ At first glance, pairing renewable energy with the oil and gas sector would seem an unlikely match. But behind the curtain, romance could be blooming.

❷ From 2000 to 2010, U.S.-based oil and gas companies invested roughly $9 billion in renewable technologies (such as wind, solar, biofuels)—roughly one fifth of the total U.S. investment in renewables over the same period. And as Canadians come to recognize that meaningful and cost-effective climate action may be the key to unlocking market access for oilsands, the appetite for an even-tighter union between these star-crossed industries could be just around the corner.

Figure 8-1 Shell launches solar- or wind-powered autonomous well platforms in North Sea

A Match Made in Heaven?

❸ In many places, the economic tide is turning in favor of renewables over fossil fuels. Wind power recently beats out new coal in Australia, and it's becoming increasingly competitive with

1. Switzer, J. (2014, April 14). *When renewables meet the oil and gas industry, opposites attract.* Retrieved from https://www.renewableenergyworld.com/articles/2014/04/when-renewables-meet-the-oil-and-gas-industry-opposites-attract.html

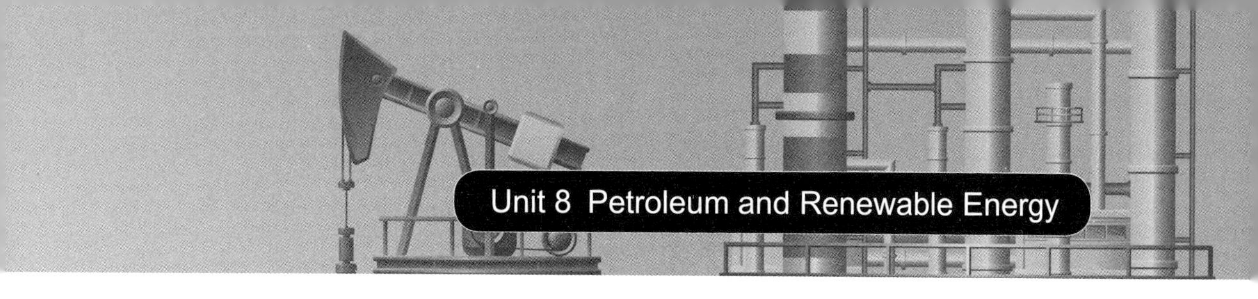

natural gas in Texas.

❹ Developing renewable energy plays to the strengths of the oil and gas sector, which include energy market insight, technology know-how, mega-project management excellence, rock-solid credit, and community engagement experience. Renewables offer a means for diversification in the face of volatile energy input costs, and a hedge against peaking oil demand in key markets. Renewable energy investments can also earn oil and gas companies favorable political capital among climate-conscious community members and decision makers.

❺ The oil and gas sector's key financial partners—institutional shareholders, banks, and insurers—are also demanding aggressive carbon management. As we discussed in a previous post, finding cost-effective or indeed profitable ways to shift carbon off the balance sheet is becoming urgent, threatening to strand reserves and cut into market valuations. Under the revised Equator Principles, for example, project developers will increasingly be obliged to demonstrate consideration of "cost-effective options" to reduce greenhouse gases as a condition for project finance. In short, oil and gas companies have both the opportunity and the imperative to invest in renewable energy.

A Volatile Courtship

❻ Oil and gas companies have been dabbling in the renewables business for a long time. In the wake of the oil shock of the 1970s, OECD governments established a range of incentives and subsidies for energy independence. This sparked a first wave of standalone renewables business ventures by the oil patch, focusing on solar, wind, and geothermal energy. Today, Chevron is the world's largest private producer of geothermal power, for example.

❼ But government matchmaking proved a fickle friend. When public incentives for renewable technologies were withdrawn in the 1980s, most companies abandoned their alternative technology investment and refocused on their core petroleum business.

❽ A second wave of oil and gas companies got into renewables in the late 1990s and early 2000s as momentum built around reaching a global climate agreement. When the global economic downturn hit and industrialized countries failed to match rhetoric with action, ventures in solar and wind by BP and Shell, among others, were sold off or quietly shuttered.

❾ Today, ethanol blending mandates in the U.S., E.U., and Canada are driving a third wave of oil and gas ventures, with Shell now the world's leading vendor of biofuels. But this wave too faces the possibility of reversal, given a growing backlash against widespread use of various biomass resources for fuel, and concerns about competition with food crops.

Couples' Therapy

❿ The chemistry between oil and gas companies and renewables ventures is likely to continue to ebb and flow with the economics and politics of the times. Yet believing these two industries have tremendous opportunity for synergy, the Pembina Institute brought a group of leading oil and gas companies together with renewable energy industry advocates last year. The objective was to better understand the dynamics at play, and to help advance the integration of renewable energy opportunities into decisions made by the oil and gas companies.

⓫ Every relationship has hang-ups, and we uncovered some significant barriers in our research and discussions, including:

- Challenging economics, stemming from low natural gas prices and the comparatively high capital costs of renewable energy, coupled with comparatively low rates of return for renewable power projects;
- A lack of national political effort on climate change and renewables, including the absence of a meaningful carbon price;
- A lack of renewable energy literacy among oil and gas engineers, encompassing both technology and current economics;
- Ad hoc and personality-driven approaches to renewable energy project development, including shifting commitment among senior corporate leadership.

⓬ Our research found that many instances where renewable energy technologies are already economic for oil and gas, particularly when competing against expensive diesel- or propane-based power in off-grid applications. For this reason, a host of on-site renewable energy "wins" remain largely untapped by the oil companies. Scaling these up could transform both sectors for the better.

Electricity Generation
- Produce electricity for within-fence operations, or sell onto electricity grid.

Heating/Cooling
- Heating and cooling for air and water, for buildings or processes.

Motive Force
- Mechanical force for applications like ventilation, cooling, pumping;
- Rotary force for applications like hammering or grinding (e.g. traditional wind mills and water wheels).

Liquid Transportation Fuels
- Ethanol and biodiesel produced from energy crops and non-food sources;
- First-generation, second-generation (ligno-cellulosic) , and third-generation (algae-derived) biofuels.

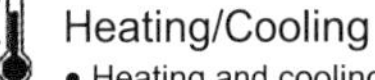

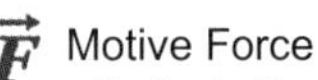

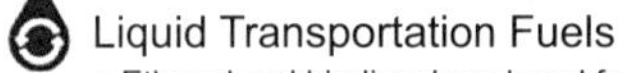

Figure 8-2 Renewable energy opportunities for oil and gas companies

Electricity Generation

⓭ As for solar for remote applications, small-scale solar photovoltaic (PV) systems have successfully been applied in remote oil and gas operations to power monitoring systems,

compressors, pipelines, and pumping stations. Companies including Cenovus, Encana, and Suncor have been using solar-powered supervisory control and data acquisition (SCADA) systems across the Canadian prairies for years, and represent the largest market for solar cells in Western Canada. In Wyoming, BP found that replacing glycol dehydrators with solar pumps paid for itself within three months.

⑭ As for renewable-powered offshore platforms, in 2002, Shell began developing offshore platforms that are fully powered by solar and wind. Today, several platforms are in use in southern North Sea gas plays, reducing operating and capital costs and improving safety.

Heating/Cooling

⑮ As for geothermal heat and pressure energy recovery, existing oil and gas wells (both operating and abandoned) connect to deep geothermal resources, meaning that many wells produce highly pressurized waste fluids and gases at temperatures as hot as 200°C. This energy can be economically recovered using existing technology. In the U.S., where there are more than two million such wells, the government is actively funding and encouraging pilot projects and mapping the available resource.

⑯ As for concentrated solar thermal for enhanced oil recovery, since many of the known heavy-oil reserves around the world have limited access to cost-effective fuel sources and are located in areas with substantial solar resources, Chevron and others are investigating concentrated solar to generate steam for enhanced recovery.

Keeping the Spark Alive

⑰ Based on our research and dialogue, the Pembina Institute believes that a stronger relationship between renewable energy and oil and gas is in the interest of both sectors and of the global climate. To fan the spark, we recommend three tonics: leadership, renewables literacy, and increased collaboration.

Leadership Mattering

⑱ In the absence of strong climate policy, the drive for renewable energy efforts in the oil and gas sector has come from internal champions within the oil companies. These champions can be empowered by a corporate renewables target, by a strong internal carbon price, and by a mandate to conduct systematic within-fence renewable energy options assessments for major projects.

Enhancing Renewable Energy Literacy

⑲ To enhance the likelihood of success, oil and gas project design engineers need state-of-the-art knowledge on renewable energy technology performance and on best practices for integrating these technologies into operations. The sector could spur innovation by bringing engineers and renewable energy technology experts together more regularly, and by launching a technology prize (e.g. XXX Prize) for successful pilots.

Increasing Intercompany Collaboration

⑳ In Canada, industry consortia such as the Canadian Oil Sands Innovation Alliance could facilitate moving sector-specific renewable energy research forward, with intellectual support from the renewable energy industry associations and non-governmental sectors, and financial support from governments. Internationally, the Society for Petroleum Engineers, and IPIECA (the Global Oil and Gas Industry Association for Environmental and Social Issues), can play crucial roles in convening technical dialogue and consolidating best practices. There is an opportunity to benchmark across the sector, to build on lessons learned across multiple companies, and to develop and adopt sector-wide guidelines for incorporating renewable options assessments in project design.

㉑ While many talk about a transition to a low carbon economy, few know how we will get there. Moving renewables from a side show into a core element of the oil and gas business could mark a tipping point, helping Canada compete in the estimated $3 trillion clean energy economy, while simultaneously providing a meaningful avenue for this major sector of our economy to address its social license challenges.

After-Class Activities

Activity One Watch the Videos About the Energy Future

Faced with the uncertainties of a potential supply crunch and the energy transition, what should oil and gas companies do? What does the future of energy mix face? Watch the following two videos and summarize the key points about the energy future.

Video 1: The Symbiotic Relationship of Fossil Fuels and Renewables

Video 2: Fossil Fuels and Renewable Power

Activity Two Practice Summary Writing

As discussed in the videos above, the predominant resources on which the global economy depends—oil, coal, and natural gas—will by no means be completely phased out of existence in decades to come. However, what about renewables? Are they going to gain the upper hand?

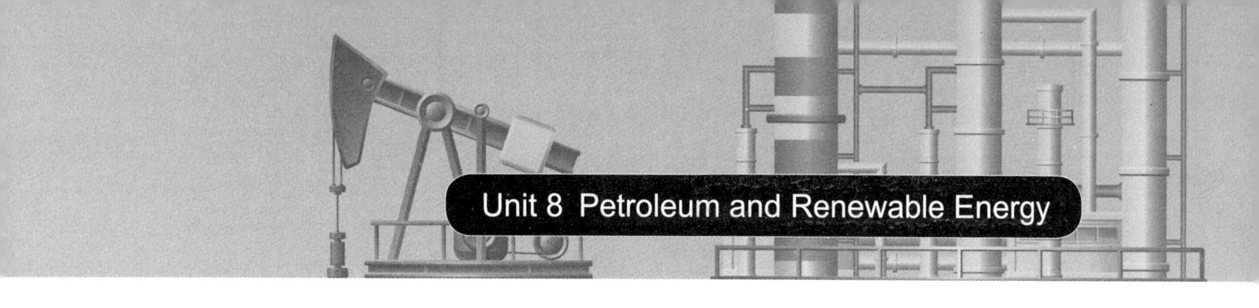

Reading C provides some insightful answers to these questions. Read it carefully and write a summary about the future energy mix with not more than 150 words.

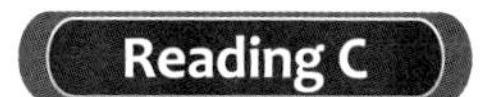

What's the Recipe for Tomorrow's Energy Mix?[1]

Benoit Laclau
EY Global Energy Leader
Andy Brogan
EY Global Oil and Gas Transaction Advisory Services Leader

❶ Over the last 50 years, hydrocarbons have become suspect. We explore the new reality shaping the future of oil and gas.

❷ For now and the foreseeable future, it's a fossil fuel world. Most of the world's energy, in fact 81% of it, starts its journey from mines and wells before being burned in vehicles and power plants. The realization that this process might not be sustainable has gained traction in fits and starts.

❸ First came scarcity and high prices with the fear that resources would eventually run out; a way of life is threatened by the taps running dry. Then, we figured out that CO_2 emissions might change the climate in a profound and irreparable way.

❹ These factors affect consumers' view of energy, the way governments regulate and tax the energy industry, and how consumers buy energy. The relevant question now is: How will the future energy mix change?

❺ We see three dimensions to the energy future—the changes in the market landscape, the choices that decision makers need to make, and the actions energy companies undertake to respond to changes. Together these factors will remake the energy industry.

Four Factors

Economic Growth

❻ The world's population and economy continue to grow. Growth in the developing world has slowed which is different from what we thought it might be, but most of the increase in economic activity in the next 50 years will be in places where the middle-class lifestyle, and the energy use

1. Laclau, B., & Brogan, A. (2019, January 23). *What's the recipe for tomorrow's energy mix?* Retrieved from https://www.ey.com/en_gl/energy-reimagined/what-is-the-recipe-for-tomorrows-energy-mix

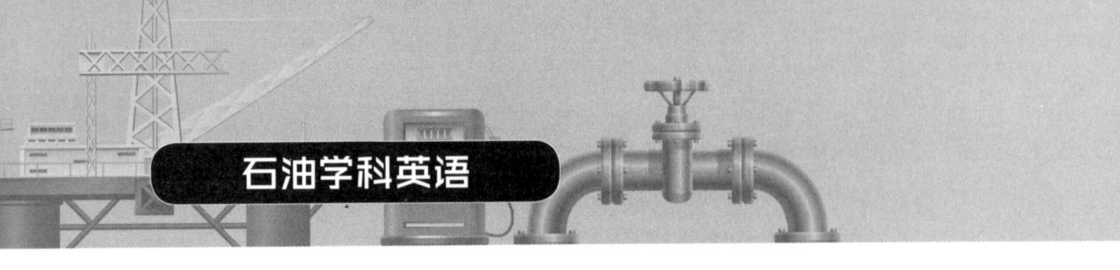

that goes with it, is a brand-new thing. Energy demand will rise at a fairly steady, predictable rate. There aren't a lot of differences between the most aggressive forecasts and the more conservative ones.

Efficiency

❼ Energy demand will rise at a rate lower than economic growth. The developed world hit an inflection point after the initial surge of new energy-using products when we started to pay attention to how much it cost to run them and the engineers focused on efficiency; vehicles went more miles on a gallon of gasoline, and air conditioners and refrigerators improved.

❽ As digitization is the next big thing, vehicles, houses, and businesses will get smarter, "driving" themselves in a way that uses the least energy and turning themselves off or powering down when it makes sense. The technology will spill over into the developing world and could completely, or at least partially, offset the wave of new energy users.

Electrification

❾ Electricity has always been and will be the vehicle for broader energy use, more energy use, and different energy use. Once there is power in a house or a business, the possibilities are endless. The biggest question today is around transportation. When will battery companies make a battery that's light enough and charges fast enough to compete with gasoline?

Decarbonization

❿ Where will more electricity come from? A growing role for coal or nuclear would be a contrarian view at this point. Advances in natural gas might price coal out of the market without regard to environmental issues. In addition, the risk of new carbon regulation, notwithstanding the current political climate in the U.S., makes long-term investments in coal plants very questionable.

⓫ Nuclear power is a carbon-free alternative, but cost and waste disposal need to be resolved. Until that happens, private capital will be scarce. If more nuclear energy is produced, it will be because of geopolitical considerations and government subsidies and guarantees. The most obvious carbon-free path is renewable power, mainly solar and wind, which has grown at a phenomenal rate in the last ten years (50% and 22%, respectively). However, those remain just a little more than 2% of the overall mix, and questions remain about cost and scalability.

⓬ Economic growth will always push the demand for energy higher, particularly when the economic growth is concentrated in parts of the world that are just learning how energy can enable a higher standard of living. Energy efficiency will continuously improve and pull against economic growth.

⓭ Energy is a scarce, costly resource, and people and business will always look to get more from less. Electrification will continue to be the vehicle for broadening and widening the use of energy. When people get access to electricity, the opportunities to use energy, including for

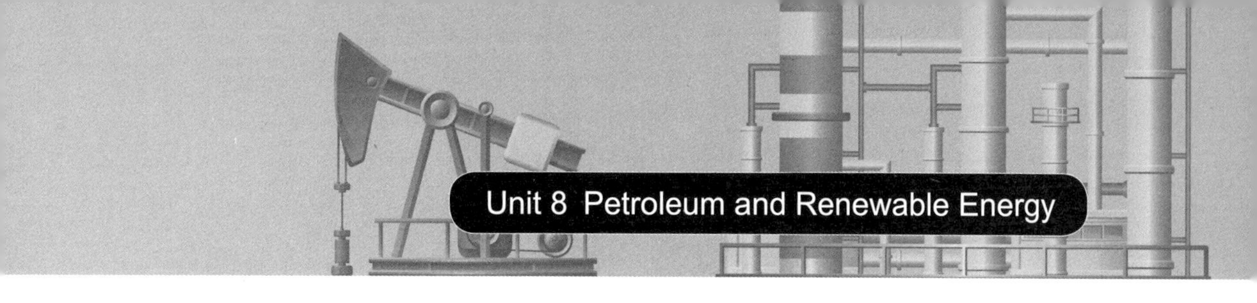

transportation, are limitless. Decarbonization is the overarching latest trend, a recognition that sustainability matters.

⑭ As we examine the drivers, it becomes clear that the decision makers will propel the energy mix from where it is to where it will be. The decision makers fall into three categories: the government and public sector, industry and technology, and consumer behavior.

⑮ We've summarized the spectrum of forecast viewpoints. Suffice to say there is no consensus. No surprise, given the number of unknowns and lack of a reliable methodology to forecast the adoption of new technologies that do essentially the same thing as the old technologies at what might be higher cost and less convenience.

How the Future Energy Mix Might Look

⑯ We've summarized and analyzed industry forecasts to create three scenarios.

Hydrocarbon Heavy

⑰ In this scenario, the energy world of tomorrow doesn't look a whole lot different from the world of today. None of the technological enablers of renewables materialize, the efficiency of oil and gas extraction continues to improve, and consumers' and governments' attitudes don't force the issue. The energy mix reflects some natural progression towards lower-carbon energy based on current trends and embedded regional commitments to renewables, but oil and gas remain dominant.

Electric Evolution

⑱ This scenario could be characterized as forced energy diversity. Renewable and electric vehicle technologies progress to a point where they're competitive at the margin, government support grows somewhat, and consumer attitudes support growing market share. Fossil fuels remain competitive and demand for oil grows in response to natural growth in population and the economy, but market share shrinks.

Renewable Revolution

⑲ In this scenario, transformation moves to at least half renewables to traditional fossil fuels. All of the variables move forward in the same direction, and almost all marginal energy use is sourced from renewables through the electricity grid. New vehicles are mostly electric by the end of the forecast, and all of the energy for those vehicles is supplied by renewable electricity. The existing stock of gasoline-burning vehicles and fossil-fired power plants exit the market via attrition while keeping oil and coal in business for a while.

⑳ While we don't know when or if each scenario might occur, choices need to be made and companies must be re-engineered to be ready for the future.

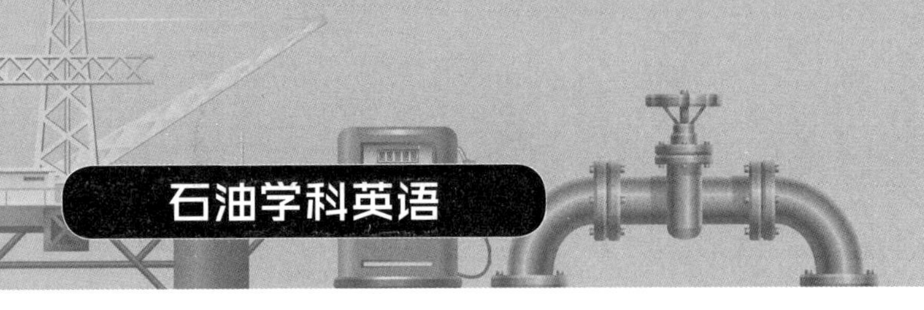

How to Get There

㉑ The energy mix of tomorrow will be different from the energy mix of today. So how do we get there, and what do we do when we get there, wherever there may be? What can energy companies do to prepare for a new and different energy future?

㉒ There's a lot to think about, but a good start to become the energy company of the future is an honest assessment of core competencies and how they fit into the new energy value chain with the biggest returns.

㉓ Core competencies are basically what a current company does best. Since their beginning, oil, gas, and power companies have been in the supply business. They know how to produce and deliver large quantities of commodity energy to consumers. To be sure, the international oil companies (IOCs) have vast networks of retail outlets, and utilities own the customer interface at the meter, but the biggest part of their value add comes before the customer drives up to the pump or the power goes through the meter.

㉔ How do those core competencies work in the new energy future? One version of that future is large-scale adoption of solar energy and lots of solar panels are likely to sit in the consumer's home and behind the meter. If companies focus on the manufacturing of solar panels and the batteries that pair with them, they may find a niche that fits within their core competencies, i.e., production and supply. Access to retail markets is essential, but it doesn't require ownership of the customer interface. A distributor model that provides access to the customer but doesn't require us to operate the retail outlet might work best.

㉕ It comes down to execution: There's a lot to accomplish, but we've distilled it to four categories.

Capital Allocation

㉖ Strategy is the intersection between what the market will want and what a company does well. There's no substitute for a little speculation on the subject of what the market will want. A bet or collection of bets will have to be made depending on a level of certainty and appetite for risk. It might include funding new technologies or expanding into new geographies. Whatever the choice is, companies must ensure they have a sustainable competitive advantage. Secondly, one shouldn't pay too much for acquired businesses, and ensure they have a solid exit strategy if the wrong bet has been placed.

Technology and Business Process

㉗ Business processes are driven by business models and it's probably best to step out into business models where existing processes can be bent, but not broken. The same statement applies to the technology chosen to support the business processes. There should be small technology steps, not giant leaps, to get the systems needed from the systems in place today.

People and Culture

㉘ Execution is all about people making the right decisions and taking the right actions while doing both of those things in the right time to make a difference. The culture of utilities and oil and gas companies tends to be deliberative. Capital discipline is rewarded; learning from failure is not. That may not work in businesses where there are a lot of turning points and the time horizon is measured in months instead of decades. Hiring, organizational structure, and incentives matter.

Brand and Reputation

㉙ Consumers today want the best, and without post-purchase dissonance. You're not the best unless customers believe that you are. Utilities are branded for regulators. Oil and gas companies are branded to guard against regulation and taxation, but price and convenience almost always overshadow reputation when we figure out where to fill up our car. To the extent that succeeding in renewable energy rests on a relationship with the government, that approach probably works. To the extent that we're trying to convince customers that we share their concern for the planet's future, it's going to be challenging to sell fossil fuels and renewables under the same brand.

Summary

㉚ An uncertain energy future leaves a wake of decisions and opportunities. Strategy choices can't really be hedged. It is time to determine what the market will want and determine the best path forward. What is manageable is the execution. It might be a long road ahead to become the most efficient, most responsive, and most respected energy player in the industry. Sounds easy to re-imagine energy, doesn't it?

Integrated Exercises

I Read the following academic words, and check whether you can use them appropriately. For those you can not, look up in a dictionary or search online about their contextual use. Write down notes to strengthen your memory.

Reading A

1. startling
2. grid
3. benchmark
4. typically
5. renowned
6. overcast
7. energy-hungry
8. momentous
9. cynicism
10. fascinating

11. tipping
12. mechanized
13. embargo
14. combat
15. dissipate
16. stride
17. staggering
18. plunge
19. dent
20. capture
21. falter
22. far-reaching
23. sustained (*adj.*)

Reading B

1. glance
2. blooming
3. cost-effective
4. mega-project
5. rock-solid
6. climate-conscious
7. insurer
8. valuation
9. imperative
10. courtship
11. dabbling
12. standalone
13. matchmaker
14. fickle
15. withdraw
16. shutter
17. ethanol
18. mandate
19. therapy
20. synergy
21. advocate
22. integration
23. hang-up
24. barrier
25. literacy
26. personality-driven
27. compressor
28. pressurize
29. substantial
30. investigate
31. empower
32. state-of-the-art
33. spur
34. consortia
35. convene
36. consolidate
37. avenue
38. mediator

Reading C

1. recipe
2. suspect
3. scarcity
4. irreparable
5. undertake
6. conservative

7. inflection
8. digitization
9. electrification
10. decarbonization
11. contrarian
12. notwithstanding
13. scarce
14. phenomenal
15. scalability
16. overarch
17. recognition
18. suffice
19. consensus
20. embedded (*adj.*)
21. marginal
22. stock
23. attrition
24. utility
25. allocation
26. intersection
27. speculation
28. incentive
29. overshadow

II Decide on the contextual meaning of the following terms and expressions.

Reading A

1. contracted cost:____________________
2. below the benchmark:____________________
3. coal-fired stations:____________________
4. solar panel manufacturer:____________________
5. the fuel mix and fleet of power-generating assets:____________________
6. tipping points:____________________
7. petrol- and diesel-engine vehicles:____________________
8. the threat of climate change:____________________
9. carbon capture:____________________
10. low-carbon energy sources:____________________
11. electric vehicles:____________________
12. the outlook for oil demand:____________________
13. large-scale policy commitments:____________________

Reading B

1. renewable technologies:______
2. cost-effective climate action:______
3. wind-powered autonomous well platforms:______
4. the economic tide:______
5. institutional shareholders:______
6. energy market insight:______
7. mega-project management excellence:______
8. oil patch:______
9. energy input costs:______
10. climate-conscious community members and decision makers:______
11. greenhouse gases:______
12. alternative technology investment:______
13. renewable energy industry advocates:______
14. renewable power projects:______
15. senior corporate leadership:______
16. off-grid application:______
17. electricity grid:______
18. electricity for within-fence operation:______
19. traditionál wind mills and water wheels:______
20. pumping stations:______
21. data acquisition systems:______
22. operating and capital costs:______
23. pilot projects:______
24. enhanced oil recovery:______
25. heavy-oil reserves:______
26. substantial solar resources:______

27. renewables literacy:______

28. internal champions within the oil companies:______

29. a corporate renewables target:______

30. a strong internal carbon price:______

31. within-fence renewable energy options assessments:______

32. renewable energy technology performance:______

33. technology prize for successful pilots:______

34. sector-specific renewable energy research:______

35. non-governmental sector:______

36. project design:______

37. a low carbon economy:______

Reading C

1. energy mix:______

2. CO_2 emissions:______

3. changes in the market landscape:______

4. economic growth:______

5. inflection point:______

6. power down:______

7. the vehicle for broader energy use:______

8. price coal out of the market:______

9. without regard to environmental issues:______

10. waste disposal:______

11. new carbon regulation:______

12. a carbon-free alternative:______

13. geopolitical considerations:______

14. government subsidies and guarantees:______

15. the government and public sector:______

16. consumer behavior:________________

17. regional commitments to renewables:________________

18. fossil-fired power plants:________________

19. an honest assessment of core competencies:________________

20. the new energy value chain:________________

21. networks of retail outlets:________________

22. a niche market:________________

23. behind the meter:________________

24. the ownership of the customer interface:________________

25. a distributor model:________________

26. capital allocation:________________

27. a solid exit strategy:________________

28. capital discipline:________________

29. the time horizon:________________

30. post-purchase dissonance:________________

III Analyze the grammatical structure of the following complex sentences, figure out the meaning of each sentence, and paraphrase them.

1. Assets such as oilfields and power plants are big investments that have operational lives lasting for many decades, and so the fuel mix and fleet of power-generating assets turn over slowly. (Reading A, Para. 9)

2. Even so, the elements are being put in place for what could be a quite sudden and far-reaching energy transition, which could be triggered by an unexpected and sustained surge in oil prices. (Reading A, Para. 20)

3. Developing renewable energy plays to the strengths of the oil and gas sector, which include energy market insight, technology know-how, mega-project management excellence, rock-solid credit, and community engagement experience. (Reading B, Para. 4)

4. When the global economic downturn hit and industrialized countries failed to match rhetoric with action, ventures in solar and wind by BP and Shell, among others, were sold off or quietly shuttered. (Reading B, Para. 8)

5. These champions can be empowered by a corporate renewables target, by a strong internal carbon price, and by a mandate to conduct systematic within-fence renewable energy options assessments for major projects. (Reading B, Para. 18)

6. There is an opportunity to benchmark across the sector, to build on lessons learned across multiple companies, and to develop and adopt sector-wide guidelines for incorporating renewable options assessments in project design. (Reading B, Para. 20)

7. Economic growth will always push the demand for energy higher, particularly when the economic growth is concentrated in parts of the world that are just learning how energy can enable a higher standard of living. (Reading C, Para. 12)

8. Suffice to say there is no consensus. No surprise, given the number of unknowns and lack of a reliable methodology to forecast the adoption of new technologies that do essentially the same thing as the old technologies at what might be higher cost and less convenience. (Reading C, Para. 15)

9. Capital discipline is rewarded; learning from failure is not. That may not work in businesses where there are a lot of turning points and the time horizon is measured in months instead of decades. Hiring, organizational structure, and incentives matter. (Reading C, Para. 28)